MODERN INFORMATION WARFARE

Operations, Doctrine and Force Structures

MODERN INFORMATION WARFARE

Operations, Doctrine and Force Structures

Col Anurag Dwivedi

(Established 1870)

UNITED SERVICE INSTITUTION OF INDIA
NEW DELHI

PENTAGON PRESS LLP

Modern Information Warfare: Operations, Doctrine and Force Structures
By Col Anurag Dwivedi

First Published in 2019

ISBN 978-93-86618-70-2

Disclaimer: Disclaimer: The views expressed in this book are those of the author and do not necessarily reflect those of the United Service Institution of India or any other organisation.

Published by
PENTAGON PRESS LLP
206, Peacock Lane, Shahpur Jat,
New Delhi-110049
Phones: 011-64706243, 26491568
Telefax: 011-26490600
email: rajan@pentagonpress.in
website: www.pentagonpress.in

Printed at Aegean Offset Printers, Greater Noida, U.P.

Contents

Foreword

Information Warfare (IW) as a term came into mainstream use post Operation Desert Storm where Information Technology played a significant role in USA inflicting a crushing defeat on the battle hardened Iraqi armed forces. In 1998 USA came out with the first Joint Doctrine for IW and India did the same in the year 2005.

The scope of IW has expanded exponentially since those early days. We now live in what is called the information age and "Information" itself is acknowledged as having become the fifth battlespace in addition to land, sea, air and space. The alleged Russian hacking of the US elections and the social media fuelled insurgency in Kashmir is a stark example of how the information battlespace can be leveraged to achieve strategic objectives. IW today is a 24/7/365 activity which is conducted by civilian and military agencies as well as non-state actors and terrorist organizations.

Despite the constantly growing relevance of IW, the underlying capabilities are extremely diverse and cover multiple unique disciplines like Electronic Warfare, Cyber Warfare, Psychological Warfare and Intelligence. These cannot be readily consolidated into unified force structures and there exist many an unresolved doctrinal dilemma. The information battlespace meanwhile continues to expand relentlessly both in terms of technology and tactics. A universal theoretical framework which can assist in analyzing the practical nuances of this new battlespace is lacking.

The book "Modern Information Warfare : Operations, Doctrine and Force Structures" by Col Anurag Dwivedi presents a de-novo theoretical model for conduct of integrated information operations and thereafter translates it into a logical and comprehensive framework for evolving operational doctrines and force structures. The book also carries exhaustive analysis of the only two existing IW force structures, namely the US Cyber Command and the Chinese Strategic Support Force to drive home the lessons.

I am delighted to congratulate Col Anurag Dwivedi and the Centre for Strategic Studies and Simulation (CS3), USI of India, New Delhi for this extremely contemporary and timely research.

New Delhi
August 2018

Lt Gen PK Singh, PVSM, AVSM (Retd)
Director, USI of India, New Delhi

Acknowledgements

I am extremely grateful to United Service Institution of India for giving me the opportunity to carry out my research under their umbrella. The intellectually stimulating environment under the inspirational leadership of Lieutenant General P K Singh, PVSM, AVSM (Retired), Director USI and the support I have received from the institution has been a major contributor to the successful completion of this work.

The Centre for Strategic Studies and Simulation (CS3) under which I carried out my research has contributed immensely towards guiding me on the right path. I am deeply indebted to Major General B K Sharma, AVSM, SM** (Retired), Deputy Director (Research) and the CS3 research coordinator Dr Roshan Khaniejo for their professional guidance.

My profound gratitude to Lieutenant General Davinder Kumar, PVSM, VSM**, ADC, former Signals Officer in Chief of the Indian Army who has been my guide for the project. He is a world renowned expert on Information Warfare and his extraordinary understanding of the nuts and bolts has helped me gain new insights and clarify many a dilemma.

Lastly, I owe a big word of thanks to my mentors and colleagues in the Information Warfare fraternity from whom I have learnt the ropes. This book is dedicated to them.

Acronyms

AI	Artificial Intelligence
ASAT	Anti-Satellite
AWACS	Airborne Early Warning and Control System
BMD	Ballistic Missile Defense
C2W	Command and Control Warfare
C4ISR	Command, Control, Communications, Computers, Intelligence, Surveillance and Reconnaissance
CI	Counter Insurgency
CNA	Computer Network Attack
CND	Computer Network Defense
CNE	Computer Network Exploitation
CMO	Civil Military Operations
CNR	Combat Net Radio
CO	Cyberspace Operations
COMINT	Communications Intelligence
COMSEC	Communications Security
CNO	Computer Network Operations
CREW	Counter RCIED Electronic Warfare
CSS	Central Security Service (USA)
DEW	Directed Energy Weapon
DCO	Defensive Cyberspace Operations
DIA	Defense Intelligence Agency (USA)
EA	Electronic Attack
ECM	Electronic Counter Measures
ECCM	Electronic Counter Counter Measures
EEP	Electronic Emission Policy
EMP	Electromagnetic Pulse
EP	Electronic Protection
EES or EMSEC	Electronic Emission Security

E-ORBAT	Electronic Order of Battle
ELINT	Electronic Intelligence
EMSO	Electromagnetic Spectrum Operations
ES	Electronic Support
ESM	Electronic Support Measures
EW	Electronic Warfare
FCC	Fleet Cyber Command (US Navy)
FININT	Financial Intelligence
GEOINT	Geospatial Intelligence
GPS	Global Positioning System
HF	High Frequency
HUMINT	Human Intelligence
IA	Information Assurance
ICT	Information and Communication Technologies
IEWS	Integrated Electronic Warfare System
IFF	Identification, Friend or Foe
IMINT	Imagery Intelligence
INSCOM	Intelligence and Security Command (US Army)
IO	Information Operations
IoT	Internet of Things
IR	Infrared
IRC	Information Related Capability
ISIS	Islamic State of Iraq and Syria
ISR	Intelligence, Surveillance and Reconnaissance
IT	Information Technology
IIO	Information Influence Operations
IPO	Information Psychological Operations
ITO	Information Technical Operations
IW	Information Warfare
KLE	Key Leader Engagement
MASINT	Measurement and Signature Intelligence

MI	Military Intelligence
MILDEC	Military Deception
MISO	Military Information Support Operations
NAVWAR	Navigation Warfare
NCW	Network Centric Warfare
NSA	National Security Agency (USA)
OCO	Offensive Cyberspace Operations
OPSEC	Operations Security
OSINT	Open Source Intelligence
PA	Public Affairs
PLA	People's Liberation Army (China)
PNT	Position, Navigation, and Timing
RCIED	Radio Controlled Improvised Explosive Device
RLV	Reusable Launch Vehicle
SATCOMN	Satellite Communications
SATNAV	Satellite Navigation
SC	Strategic Communication
SEAD	Suppression of Enemy Air Defense
SIGINT	Signals Intelligence
SSF	Strategic Support Force (China)
STO	Special Technical Operations
TBA	Tactical Battle Area
TECHINT	Technical Intelligence
TRB	Technical Reconnaissance Bureau (China)
UAV	Unmanned Aerial Vehicle
UCAV	Unmanned Combat Aerial Vehicle
USASOC	US Army Special Operations Command
UHF	Ultra High Frequency
VHF	Very High Frequency
VLF	Very Low Frequency
WHAM	Winning Hearts and Minds

SECTION 1

IW Theory

CHAPTER 1

Emergence of Information as a Battlespace

Historical Perspective

Information has been an integral part of warfare since time immemorial. "The Art of War" written in 6th century BC by Sun Tzu and Chanakya's equally well known treatise "Arthashastra" written in 3rd century BC lay great emphasis on the use of deception and psychological (psy) warfare as well as the employment of spies to gain information (intelligence). The famous Mongol warrior Genghis Khan (1162-1227 AD) was believed to send covert agents in advance of his attacks to carry out psy warfare by narrating the brutality and invincible size of the Mongol army. He also deceived his enemies by using feigned retreats luring them into ambushes, and targeted enemy messengers to prevent commanders from communicating with each other[1, 2]. The ways in which information can be exploited to achieve war objectives has kept evolving with time.

Transmission of information on the battlefield entered a new dimension after the invention of wireless in late 19th century. Signals Intelligence (SIGINT) soon emerged as a new discipline to intercept these wireless transmissions. The discipline evolved further during World War-I with the use of deceptive broadcasts and jammers to disrupt enemy command and control. Invention of the Radar during World War-II and the advent of air warfare led to passive countermeasures like use of chaff, and active countermeasures like radar jammers. These measures and countermeasures collectively came to be described as Electronic Warfare (EW). In recent years Cyber Warfare has emerged as yet another major discipline which is directly related to information and information systems.

The common unifying thread amongst these varied disciplines is that they all manipulate information or information based systems and processes to influence the outcome of war. Organized, synergistic and collective application of these diverse information related capabilities can be termed as Information Warfare (IW).

IW is however much more than a mere grouping of disciplines (or capabilities) which share information as a commonality. "Information" is now also deemed to be a distinct battlespace just like land, sea and air. We therefore need to understand what exactly a battlespace is and why information has become one. Doing so will also help us reconcile as to why it is only in recent years that we have grouped the diverse information related capabilities into a distinct class of warfare called IW.

The Evolution of Warfare

While War has remained a near constant of human existence, the methods of waging war have constantly evolved because of innovation. Understood and applied correctly in terms of strategy and operational doctrine, innovation has invariably surprised the enemy and proved decisive towards determining the outcome of conflicts.

Not all innovations have a technological foundation. Guerilla Warfare for example was a tactical innovation whose history can be traced back to around 600 BC[3]. Terrorism and insurgency are merely its modern avatars. Introduction of "war elephants" by ancient Indians and their subsequent adoption by the Persians and Alexander around 330 BC was similarly a tactical innovation. That said, the most transformational changes brought about in war fighting, at least in modern times, can be attributed to scientific and technological innovation. Changes to operational doctrine and tactics have followed suit. The invention of Submarines (1863), Airplane (1903), Tanks (1915) and Nuclear Weapons (1945) are notable examples where scientific and technological innovations resulted in a profound impact on operational doctrine and warfighting strategy.

It must also be remembered that neither of the aforementioned scientific or technological innovations were abrupt developments. Each one of them took years of refinement before they attained critical mass and were effectively employed in war. It took several more years before operational doctrines and tactics matured to the extent of becoming common parlance, practice and part of training. The first tanks for example were employed during World War-I but they fully came into reckoning only during World War-II and the new operational doctrine of Blitzkrieg came into being. The doctrine of Mutually Assured Destruction (MAD) likewise matured much after the first nuclear bombs were dropped on Japan in 1945.

An instructive example is that of the Machine Gun. Machine Guns underwent evolution and development for over 300 years before technology reached an inflection point and the US Army issued four machine guns per regiment in 1912. This increased to 36 machine guns per regiment by 1919 – a rapid nine fold increase[4]. Even more dramatic is the case of fighter aircraft. The first successful controlled flight of a fixed wing aircraft was demonstrated by Wright brothers in December 1903. Less than two decades later more than 50,000 airmen had died in combat during World War-I[5]. Three decades later, World War-II saw the most extensive use of airpower in the history of warfare and "Air" came into reckoning as a decisive battlespace. Similar is the story of submarines, the tank and nuclear weapons. Technology got rapidly inducted once it crossed an inflection point and thereafter played a pivotal role in all strategic calculations and battles. Tactics, doctrine, equipment and strategy were suitably renewed.

A key indicator of technology induced doctrinal shift is that it results in a manifest change in the equipment holding as well as organizational profile of the Armed Forces. Secondly, it results in creation of new manpower specialties that hitherto fore did not exist, viz. fighter pilots. Lastly, doctrinal shift is visible in the form of rapid adoption of offensive and defensive counter-measures viz. Anti-Submarine, Anti-Aircraft and Anti-Tank warfare. In rare cases the innovation is so transformational that it leads to creation of a new battlespace. Aircrafts were one such great innovation.

In the modern era semiconductors, digital communications, computers and networking have ushered in a similar transformation and they are commonly referred to as Information and Communication Technologies (ICT). ICT as an innovation followed a trajectory of rapid adoption from 1980s onwards when it crossed an inflection point. Numerous old weapon platforms and equipment got replaced by a new generation of equipment that were digitally controlled and embedded with microprocessors, sensors and communication. The main virtue of this ICT enabled generation of battle platforms was their ability to generate, integrate and assimilate information thus enabling synchronized application of combat power beyond visual range. The resultant new doctrine is commonly known as "Network Centric Warfare (NCW)". The resultant battlefield is often termed as the "Digital Battlefield".

NCW is slightly difficult to comprehend because it is not uniquely identifiable with any specific weapon platform. It instead relies on creation of a force multiplier effect by synchronizing combat power of geographically dispersed forces through shared situational awareness and better command and control. A scaled down example that can help understand this concept is the famous doctrine of Blitzkrieg. Blitzkrieg was to a large extent attributable to a simple innovation by Heinz Guderian of fitting each German tank with a wireless set [6]. The resultant situational awareness and improved command and control enabled the German Panzer Divisions to extract far greater combat potential from their tanks and outmaneuver adversaries having similar platforms.

The Information Age

While the Internet came into existence earlier, it was actually only around the period 1995-2000 (depending on the geographical region), that it reached an inflection point and started becoming a household thing. Two things prompted this. Firstly computers (processors, memory storage and associated hardware) became affordable and turned into a household gadget. Secondly fiber optic and satellite communications technology matured and got widely deployed thereby allowing networks across the globe to interconnect over high bandwidth links. Once telephone companies started offering last mile connectivity, computers could then plug into this global grid called the Internet or World Wide Web and become a part of it. Web applications grew exponentially and so did the number of users.

I still remember buying my first desktop computer in 1998 from Calcutta for a princely sum of forty thousand rupees (roughly five month's salary). It had a processor running at 166 MHz clock speed and the hard disk was 2 Giga Bytes. I had to convince the vendor to upgrade the RAM from 32 Mega Bytes to 64 Mega Bytes which he felt was a complete waste of money. I got Internet in the year 2000 and it was a 33.6 Kilobits/second dial up connection coming over the telephone line.

Two decades later I now have a computer which costs less than a month's salary. It has a 2.8 GHz processor, a 1 Tera Byte hard disk and 8 Giga Bytes of RAM. The Internet connection is 100 Megabits/second and comes over a fiber optic cable. Not just the memory and data rates, but also the type of computing devices (viz. smart phones), the applications and services have changed dramatically. Leaders, government departments, businesses, services, infrastructure and individuals are all now inextricably linked to the Internet. There are also an exponentially increasing number of non-human sensors, machines and intelligent software programs joining this ecosystem. So profound are these changes that we refer to this new era as the "Information Age". Everyone and everything is riding the information grid and it has transformed every conceivable aspect of human civilization including warfare.

Figure 1.1: Growth of Internet Users (1996-2018)

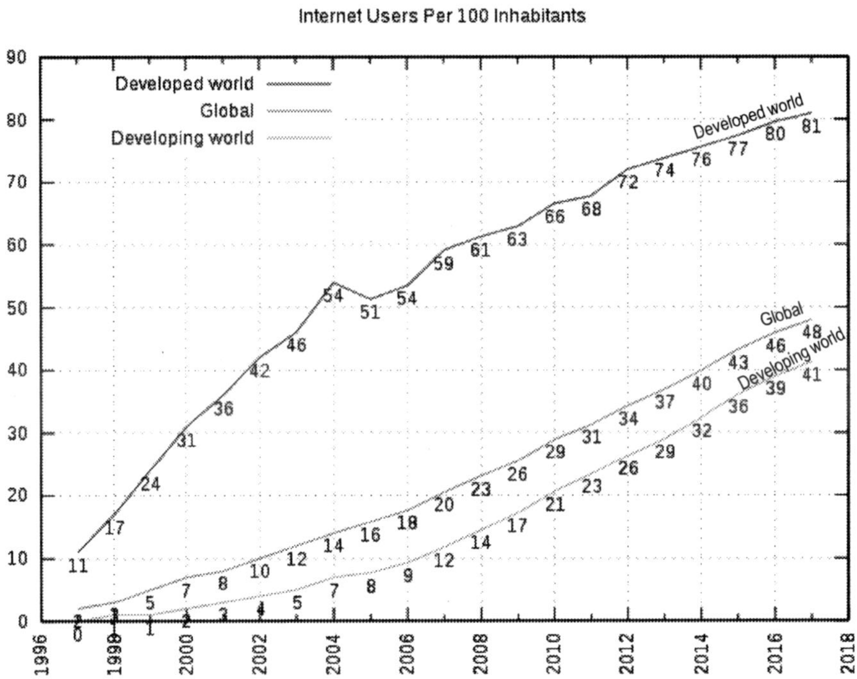

(Jeff Ogden (W163) and Jim Scarborough (Ke4roh) (https://commons.wikimedia.org/wiki/File:Internet_users_per_100_inhabitants_ITU.svg), "Internet users per 100 inhabitants ITU", https://creativecommons.org/licenses/by-sa/3.0/legalcode

Above preamble of NCW, the Digital Battlefield and the Information Age is essential to place information as a battlespace in its proper context. Bits and bytes are the bullets of this new battlefield and just like manufacture and firing of bullets would have rapidly multiplied coinciding with proliferation of machine guns; similarly the quantum of information generated (manufactured) and transmitted (fired) exponentially multiplied with the advent of NCW and the Information Age. Even the humble Infantry soldier is now envisaged to be equipped with sensors and networking capability in order to be part of a common information grid. The Information Technology (IT) specialist is an indispensable new type of warrior found in every military organization and their numbers are constantly growing.

Information as the Fifth Domain and a Battlespace

While every major innovation has led to doctrinal changes, very few have led to creation of new battlespaces. "Land" and "Sea" were the two ancient battlespaces where humans waged war. These were joined by "Air" during the World Wars. From 1970s onwards "Space" became the fourth battlespace witnessing a keen contest. The latest addition to this is "Information" and it begs asking as to why it must be given such an exalted status especially when information has always been a part of warfare.

NCW, the Digital Battlefield and the Information Age have undoubtedly induced a doctrinal shift, resulted in a change to equipment and organizations, creation of new manpower specialties and also adoption of offensive and defensive countermeasures; but similar has been the case of other major innovations like submarines and tanks as well. This alone is not sufficient to start calling "information" as the fifth battlespace.

The reason we call information as a battlespace is because it fulfills three fundamental attributes of a battlespace. Firstly, it is a distinct and unique domain with independent characteristics. Thus campaigns can be planned and firepower, maneuver, control, agility and awareness can be applied within it independent of the operations taking place in other battlespaces. Secondly, a battlespace is one which decisively shapes the outcome of battles and can help achieve strategic objectives in combination with the other battlespaces or even single handedly by itself. Thirdly, a battlespace lends itself to multiple types of operations like attack and defence. Information meets all these criteria just like land, sea, air and space.

Air did not become a battlespace the moment Wright brothers flew their prototype. It became a battlespace only after it started to have a decisive impact on the outcome of wars. Similarly, it is only in today's Information Age that "information" can independently alter the outcome of conflicts and make a decisive strategic impact. This is the reason why "information" has been deemed as a battlespace and IW has emerged as a distinct new way of warfighting only since around year 2000 and not earlier.

The table below tries to summarize some key innovations and the resultant new types of warfare and battlespaces that came into being as a result:

Innovation	*New Type of Warfare*	*New Battlespace and Approximate Year of Coming into Existence*
-	Land Warfare	Land (Ancient)
Ships	Naval Warfare	Sea (Ancient)
Submarine	Submarine Warfare	X
Tank	Tank Warfare	X
Aircraft	Air Warfare	Air (Post 1915*)
Atom Bomb	Nuclear Warfare	X
Spacecraft	Space Warfare	Space (Post 1970*)
ICT	NCW	X
ICT + Internet	Information Warfare	Information (Post 2000*)

* Approximate year

It is interesting to note that submarines, tanks, fighter aircrafts and nuclear weapons did not result in the older types of warfare going obsolete. Over a short cycle lasting several decades multiple transformational innovations co-exist and get employed in complementary fashion even though the new one may gain increasing prominence. Operational doctrine gets overhauled after each such transformational induction and tactics and equipment design are suitably altered to adapt to the new paradigm. However when we look at longer time cycles, the war elephants as also the horse cavalry and archers actually did become obsolete. The new supersedes the old but gradually.

As already brought out, a battlespace has certain unique characteristics which is why land warfare, sea warfare and air warfare are each distinct from each other though complementary. When it comes to Information as a battlespace, the uniqueness lies in its all-pervasive, ubiquitous and virtual nature. Unlike battleships and fighter jets, the Armed Forces do not have complete monopoly over production, possession or operational employment of information capabilities. Nor is there a distinct dividing line between military and civilian information and they both converge as we move from tactical to the strategic level. Information is thus a complex all-encompassing battlespace open not just to the armed forces but also to government departments, politicians, businesses, hackers, geeks, trolls, analysts, authors, celebrities, non-state actors as well as terrorists and even bots. Each of them acts as an information warrior trying to leverage the battlespace to achieve their respective objective. This complex collusive environment comprising of various actors, sources and types of information, as well as the underlying storage, transmission, acquisition and analytical mechanisms is the information battlespace or the "infosphere". Technologies both old (verbal communication, printed material, radio, TV) and new (networks, websites, social media, apps) coexist within this battlespace. New ones like Artificial Intelligence (AI) are stepping in. The resultant warfare in this diverse, ever expanding battlespace is IW.

Also as mentioned earlier, every major innovation resulted in a rapid adoption of offensive and defensive countermeasures. The advent of tanks for example resulted in adoption of anti-tank weapons (offensive) and ditch cum bunds (defensive). Anti-submarine and anti-aircraft measures likewise became specialized disciplines in themselves. Once "information" became the enabler of network-centric war, it was but logical that anti-information measures would also come into existence. IW can therefore be viewed both as a logical progression of the NCW doctrine as well as a consolidation of the counter-measures against it.

Finally, like any other battlespace – there is a struggle for dominance or superiority that takes place within the information battlespace. Just like ships and submarines battle it out at sea or dogfights that happen in aerial combat – similarly both offensive and defensive operations are conducted within the infosphere to gain decisive information superiority or at least a favorable information edge. The presence of many types of warriors and their varied objectives complicates this battle. This war involving "information" and which takes place in the infosphere is IW. Our focus shall be on the military aspects.

In its infancy, IW and its primary affiliated domains of EW, Cyber and Psy Warfare were used much like preparatory bombing i.e. to destabilize the enemy before the tanks and infantry rolled in. We are now witnessing a shift wherein information operations by themselves achieve strategic objectives without relying on armed conflict. This has a profound bearing on operational doctrines and force structures. Sun Tzu had famously said "To win one hundred victories in one hundred battles is not the acme of skill. To subdue the enemy without fighting is the acme of skill". Such is the growing power of IW.

IW is tangibly different form earlier revolutions in military affairs. IW rarely involves any direct transaction of kinetic energy or fire. The intangible nature of IW vis-à-vis kinetic weapons can be understood by using the analogy of digital currency and printed banknotes. If the aim is to achieve a financial transaction, both forms can achieve the objective. Also, while physical cash is more useful for buying street food, the digital version is more effective for online purchase of an air ticket. Hence IW and Kinetic weapons co-exist and complement each other. As Deng Xiaoping said – "It doesn't matter whether a cat is white or black, as long as it catches mice."

IW in Civil and Military Domain

Consider a hypothetical example wherein the strategic war objective is to bring about a favorable regime change in another country. The kinetic method was Operation Iraqi Freedom which replaced Saddam Hussain. Its IW equivalent is the alleged hacking of US Presidential election by Russians[7]. Both achieved a favorable regime change and the later seems to have achieved it which much greater sophistication and deniability (it is the Americans who are blaming themselves and fighting amongst each other) and at negligible cost. And whereas the Americans are debating whether Russians did it or not, there are no debates on whether such strategic regime changes can actually be inflicted through the information battlespace or not. It certainly can be. Similarly if the stated aim is to destabilize a region – the social media fuelled insurgency has recently achieved in Kashmir what infiltration and cross-border support from Pakistan was failing to accomplish for several years[8]. The Islamic State of Iraq and Syria (ISIS) is yet another example of how the information battlespace is being leveraged to wage a new kind of warfare on a global scale. Some call it Fourth Generation Warfare but it is essentially about information operations. Indoctrination, motivation, deterrence, intimidation, force projection, training and funding have all moved to the virtual information battlespace. The new way is gradually superseding the old.

It is also clear from the above examples that modern IW straddles both military and non-military domains and considerable overlap exists particularly in the area of cyber warfare, psy warfare and deception. EW on the other hand comes into play mostly during war or counter-insurgency and remains mostly a military area of activity with considerably less application in the civilian domain. We shall go into details later.

Also evident is the fact that IW is being conducted both during peace and war. Covert civilian and non-state IW predominates during peace while overt military IW operations come alive in a major way during war and counter insurgency. It would however be incorrect to assume that military IW is totally dormant during peace. Training, profiling of enemy IW targets and capability building are vital peacetime activities without which no meaningful information operations can be conducted during war.

While the above description of information as a new battlespace and IW as the new type of warfare gives us a fair understanding; we still need a sound theoretical basis to comprehend its nuances and evolve tactics, operational doctrine and force structures. The next chapter attempts to lay down such a framework.

ENDNOTES

1. Wikipedia article "*Mongol military tactics and organization*", Web URL: https://en.wikipedia.org/wiki/Mongol_military_tactics_and_ organization# Psychological_warfare_and_ deception accessed on 01 March 2017
2. The Information Warfare Site, "*Psychological Operations*", Web URL: http://www.iwar.org.uk/iwar/resources/iw-course/IW110%20Reading.doc accessed on 08 Mar 2017
3. Leonard, Thomas M, Encyclopedia of the Developing World, 1989, p. 728.
4. Wikipedia article "*Machine gun*", Web URL: https://en.wikipedia.org/wiki/Machine_gun#History accessed on 20 Jan 2017.
5. John Buckley, "*Air Power in the Age of Total War*", (Taylor & Francis, 1998), p.43.
6. Heinz Guderian, "*Panzer Leader*", (Da Capo Press, New York, 2001, originally published in 1952), p.20.
7. Massimo Calabresi, "*Hacking the Voter*", (Time Magazine, 10 Oct 2016).
8. Thakur Sankarshan, "*Making of a social-media militant*", (The Telegraph, 11 July 2016).

CHAPTER 2

The Information Terrain

Let us forget the traditional IW capabilities of EW, Cyber and Psy Warfare for some time and instead try to analyse the information battlespace based on first principles.

Consider the exploitation of social media by terrorist organizations like ISIS or Lashkar-e-Taiba versus the government agencies which contest their propaganda and ideological indoctrination. Contrast this with a piece of ground (or air space or sea lanes) being contested by two opposing forces each trying to capture dominating positions and deny freedom of manoeuvre to their enemy. Clearly the information battlespace is very similar to the traditional battlespaces. Where we go wrong is in equating Facebook, Twitter and WhatsApp etc with landmarks to be occupied or dominated – they are not. These are merely systems or platforms currently manifest in the information battlespace and which would be obsolete in a few decades. Let us elaborate upon this further.

The area of Gaza has seen armies of Alexander (332 BC), Napoleon (1799) and the Israelis (2008-09) in combat. Over these centuries, horse cavalry and archers have been replaced by tanks and machine guns yet the contours on the maps carried by Alexandrian, Napoleonic and Israeli commanders have hardly changed. The horse cavalry and tanks are not landmarks; they are merely the technologies or platforms treading the same piece of ground at different points in time.

Every soldier understands the importance of ground and terrain. For example when we talk of land as a battlespace, the first and foremost term of reference is whether one is fighting in plains, mountains or desert. Every other consideration including operational plans, force structure, tactics and equipment is subordinate to it. We do not take up linear defense in deserts or attack with a tank division in mountains because the terrain is not conducive to such operations.

It is due to this enormous importance of terrain that formulation of every plan be it attack or defense, always starts with a terrain analysis. Terrain analysis is typically done under the subheads of Observation, Cover, Obstacles, Key

Terrain Features, and Approaches etc. Once completed it leads to a set of viable options and comparison of these options gives us the final operational plan. When it comes to information as a battlespace, we lack such an analytical framework.

Mapping Terrain in the Information Battlespace

Clearly, there is a need to classify the information battlespace into distinct types of terrain and identify underlying characteristics which facilitate (or discourage) a particular class of information operations. History provides us some valuable clues.

The terrain which we see today has evolved over millions of years. Earth initially was simply a ball of hot gases which cooled down resulting in solid rock. These solid land masses thereafter collided with each other and in combination with other geological phenomenon resulted in the present landscape comprising of the atmosphere, oceans, rivers, plains, plateaus, deserts, jungles and mountains. The information landscape has similarly evolved over time. Let us go into more detail.

The ancient Indian epic Mahabharat (10th Century BC) describes how the warrior Abhimanyu learnt to get inside the most confounding defensive formation called Chakravyuh by hearing his father Arjun narrate it to his mother Subhadra. The transmission and acquisition of information was entirely in the cognitive realm. Pheidippides who ran from Marathon to Athens and conveyed information about victory in the Battle of Marathon (490 BC) similarly carried the information in his mind and conveyed it via spoken words. What these famous examples tell us is that data and information in early times existed predominantly in cognitive form and was passed on from person to person through word of mouth (ie cognitive assimilation). Physical information existed only to a very limited extent in the form of handwritten papyrus, parchments and engravings. Electronic information did not exist at all. Both the volume of information as well as the velocity of its transmission was low.

Evolution in printing techniques thereafter resulted in large volume of abstract cognitive information getting committed to physical form. As a result, both the volume of information and the velocity of its transmission increased by several orders of magnitude. The invention of radio and telegraphy added further velocity to the flow of information and also led to the creation of electronic information (albeit transient in nature). Each of these technologies were perhaps just as revolutionary in their times as computers and Internet appear today.

In our present era computers, ICT and Internet have multiplied the volume, variety and velocity of information exponentially. Also transient electronic information has been supplemented by permanent electronic storage. The doctrine of NCW for example is essentially based on leveraging the volume and velocity of electronic information. The countermeasures are likewise aimed at negating this leverage.

Figure 2.1: The Changing Nature of Information

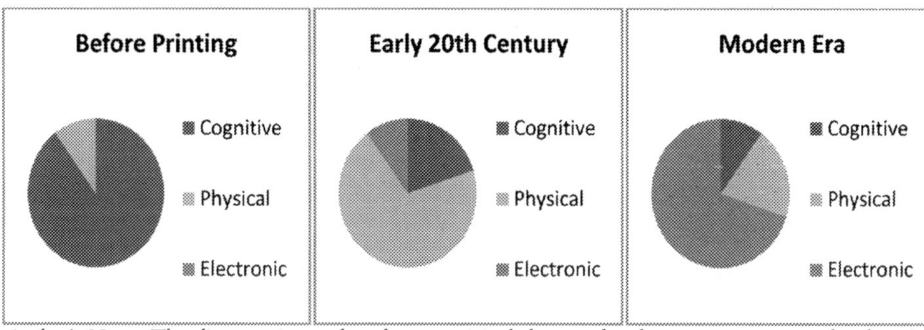

Author's Note: The diagram is not based on empirical data and only meant to convey the changing nature of information

Figure 2.2. The Changing Volume and Velocity of Information[1]

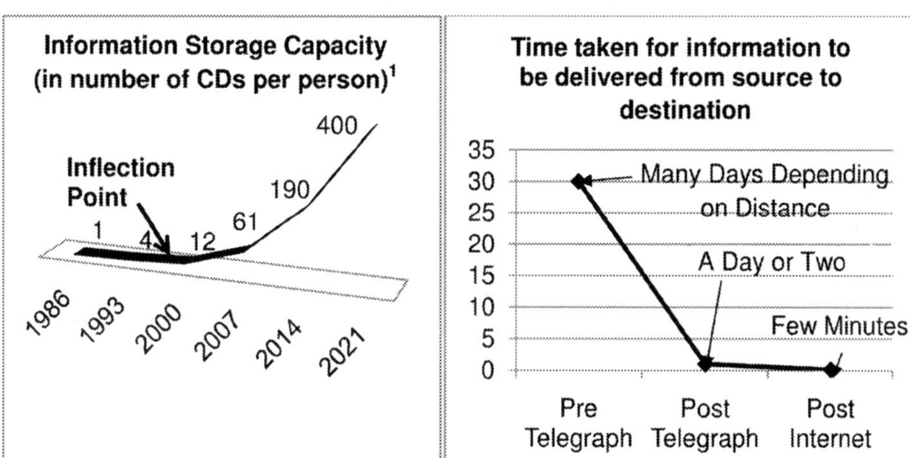

Author's Note: The diagrams above are based on extrapolation and experience and not entirely on empirical data

What is evident from the above discussion is that the present information landscape consists of three distinct types of information namely cognitive, physical and electronic. Alternatively, we can also say that the information landscape comprises of three types of terrain – namely cognitive, physical and electronic. Thus a Top Secret letter is a piece of physical information as well as an objective (or target) in the physical information terrain which maybe attacked (or needs to be defended). The terrain itself is a group of objectives having similar attributes. An analogy would be a contiguous group of high peaks which qualifies to be called as mountainous terrain. The peaks are objectives having targets as also collectively the terrain.

The next obvious question is how will this hypothetical classification into cognitive, physical and electronic terrain help us in conduct of operations within the information battlespace? Also what are the key attributes of these information terrains and can we carry out a detailed information terrain analysis similar to what we do in case of conventional battlespaces? The "information cycle" helps us take our discussion forward and answer these important questions.

The Information Cycle

Philosophers have asked whether a tree falling in a forest has actually made a sound if there was no one to hear it. A more practical analogy would be of an ultrasound whistle which can be heard by a dog but not a human. Capturing the data is therefore the start point without which information does not exist. This is achieved using sensors which can be of many types viz biological, chemical, mechanical, electric, magnetic, optical, or a combination.

When this sensor data is analysed by a processor, it yields information. Processors can be biological (brain) or electronic (semiconductor chips + software algorithms). Both data as well as information can be stored, retrieved and retransmitted. Accuracy of data capture, transmission losses and efficiency of the analytical algorithm (processor) plays a role in deciding the quality of information.

Once we have generated information, it can then also be synthesized and correlated with other pre-existing information to yield knowledge. Individual ability and experience come into play when synthesizing knowledge from information. Wisdom is a still higher form.

This entire process which starts from acquisition of data by a sensor and culminating into knowledge is called the information cycle. The information cycle can be diagrammatically depicted as under:

Figure 2.3: The Information Cycle

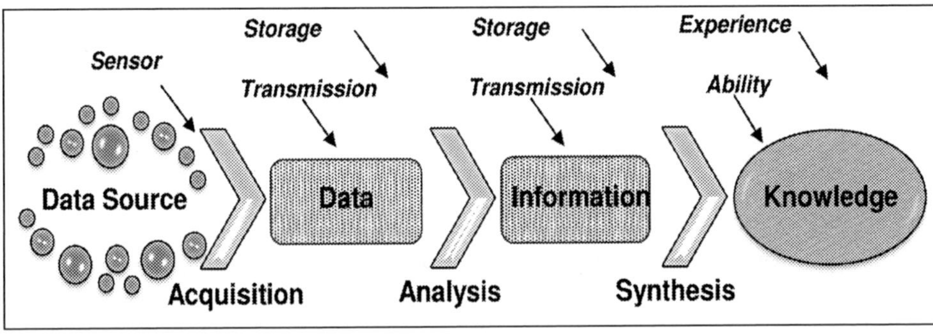

For the purpose of understanding the information terrain we will restrict ourselves to the first two steps leading to information and omit the higher abstract manifestations of knowledge and resultant wisdom.

One look at the above diagram makes it clear that data source, sensor, acquisition, transmission, storage and analysis are fundamental to the information cycle and decide the nature and quality of the data/information which is generated.

Information Terrain Analysis

Just as we do terrain analysis in the land battlespace under the heads of observation, cover, obstacles, key terrain features and approaches; a similar terrain analysis can be done in the information battlespace based on the constituents of the information cycle ie data source, sensors, acquisition, type of data transmission, analytical processing and storage mechanisms. Done correctly, it immediately yields the best approach (or combination of approaches) to attack the enemy's information or the best defensive design to protect one's own information. It also allows offensive and defensive resources to be specifically identified and tasked in support of the overall operation plan. Subsequent chapters will clarify what these operations and resources are in more detail.

Being a new concept, let us try to understand how to carry out an information terrain analysis and derive options and plans with the help of some practical examples. The examples also help us gain insight into some unique attributes of the information battlespace and how these attributes in turn modulate the operations.

Example 1. A patrol commander observes clouds overhead and concludes it will rain. Here the data source is a natural object, eyes act as sensors, the data created is in the form of sensory impulses and it is transmitted using the nervous system. The brain does the analytical processing to classify observed clouds as low, dark, dense and thereafter uses this information in conjunction with past experience to synthesize it into knowledge that it will rain. The brain also provides the required storage space (memory) where the information is stored. As may be seen, the entire information cycle as shown in Figure 2.3 has been played out in the cognitive domain. The information terrain analysis can be done as under:

Data Source	*Sensor and Acquisition*	*Data Generated*	*Transmission*	*Analytical Processing*	*Storage*
Natural object	Biological (Cognitive)	Sensory (Cognitive)	Neural (Cognitive)	Neural (Cognitive)	Neural (Cognitive)

Since all key elements (targets) lie in the cognitive terrain, operations and weapons applicable to electronic and physical terrain are ruled out (just like we ruled out tank warfare in mountains). Even cognitive attacks are not possible because the senses, thinking process or memories are all internal and inaccessible.

The reason for giving this example is to highlight that analytical processing and storage are inaccessible in case of cognitive information cycles. Cognitive information can only be acquired if transmission is external (viz verbal communication or gestures) but not if it is internal (neural).

The only way to attack cognitive information cycles is to either deny or degrade the acquisition of data or to inject data at the source or sensor level. Had the patrol commander observed an enemy machine gun position it would have been possible to inject misleading information through camouflage and dummy positions but here the data source is a physical phenomenon (or natural object) which cannot be manipulated. IW is mostly inapplicable.

While capability and disposition can be discerned by acquiring physical and electronic information, the intent mostly lies in the cognitive realm and it is extremely difficult to discern or manipulate. Similar examples can be thought of in the non-military realm too.

Example 2. A military commander goes for field reconnaissance, makes mental and written notes and comes back to the headquarters for further brainstorming. Detailed verbal discussions over paper maps are followed by an operational order being typed out on a manual typewriter (or handwritten) and sent to subordinate units through a dispatch rider. Once again the entire cycle has been completed and the data/information has only existed in the cognitive or physical form (paper documents). No electronic data or information has been created. The information terrain analysis would look as under:

Data Source	*Sensor and Acquisition*	*Data Generated*	*Transmission*	*Analytical Processing*	*Storage*
Natural objects	Biological (Cognitive)	Sensory (Cognitive) and paper documents (Physical)	Verbal discussions (Cognitive) and paper (Physical)	Neural (Cognitive) and maps (Physical)	Neural (Cognitive) and paper (Physical)

Since most of the elements (targets) lie in either the cognitive or physical terrain, operations of the electronic type do not come into play. However, the cognitive element here is different from the earlier example because a verbal transmission is taking place whereas in the earlier example it was purely internal (neural) thinking. Also physical information in the form of documents is being generated. Espionage and sabotage are therefore possible. The information cycle can be attacked by bugging the headquarters, planting a mole or by ambushing/subverting the dispatch rider carrying the operational order. If one were a defender, one knows the enemy's most likely approach and how to defend against such attacks.

An important aspect to note is that acquisition of physical information often involves use of espionage (i.e. agents or eavesdropping devices). An attack on physical information similarly involves physically intercepting the target either covertly or by force. Defense of physical information likewise involves preventing unauthorized access through physical protection. We observe that operations involving physical information may spillover into conventional battlespaces.

Example 3. A Radar scans the airspace by firing electromagnetic pulses. The data received in the form of reflected pulses is sensed by a receiver and fed into the radar electronics (hardware and software) which analyses them to generate information about objects present in the airspace and discriminates amongst the objects. Unidentified intruders are reported to interceptor aircraft and air defense units. Here the data source is a manmade object (aircraft), the sensor is electronic, transmission takes place as electromagnetic radiation and analysis is carried out in the electronic as well as cognitive domain (by the radar operator). The information terrain can be analysed as under:

Data Source	*Sensor and Acquisition*	*Data Generated*	*Transmission*	*Analytical Processing*	*Storage*
Man-made object	Electronic	Electronic	Electronic (radiating)	Electronic (hardware + software) and Cognitive	Electronic

Here the information cycle predominantly plays out in the electronic terrain. The radar can be picked up by electronic systems and the receiver can be overloaded (jammed) or fed false electronic pulses (electronic deception). It is also possible to corrupt the software program residing in the radar electronics if it can somehow be accessed and manipulated. Cognitive operations directed against the radar operator are difficult to execute. The best approach therefore is to employ weapons and tactics suitable for the electronic terrain. Both the attacker and defender can make their operational plan accordingly.

Example 4. A military formation uses a Wide Area Network for its operational communications. The network comprises of a number of interconnected nodes (essentially switches or routers). Voice and data services are provided to headquarters and units in vicinity of the nodes using last mile links. For security reasons, this network does not use any radiating media, communications are end-to-end encrypted, the terminals only run proprietary applications and only earmarked personnel can access the terminals. It is also not connected to the Internet. In this particular example there are no sensors, instead the data is being generated by human operators. The storage and transmission are electronic. The analytical process (switching and routing) underlying this network are also electronic (hardware and software). The underlying information terrain can be analysed as under:

Data Source	*Sensor and Acquisition*	*Data Generated*	*Transmission*	*Analytical Processing*	*Storage*
Human	Electronic	Electronic	Electronic (non-radiating)	Electronic (Hardware + software based switching and routing)	Electronic

Once again since the information cycle is predominantly playing out in the electronic domain, cognitive terrain operations are ruled out. Also the transmission is non-radiating and it cannot be jammed from a remote location. However, physical information operations of tapping the cables and damaging them through sabotage are possible. Electronic attacks against the switches and routers are also possible if we can somehow gain access to them through sabotage or by inducing an intentional or accidental attack through one of the human operators. This is a very peculiar result wherein the targets are in the electronic terrain but the attack and defense options are similar to the physical realm of information. Similar analysis helps the defender (in this case the network administrator) to identify the enemy's most likely approach.

Example 5. A military formation deploys an array of surveillance radars and sensors for monitoring the forward zone. The sensors relay data over a network of radiating and non-radiating media to a surveillance centre. There are also human patrols and a network of observation posts which send back data. The entire data is fused and analysed using software (viz GIS) and hardware (computers) to produce situational awareness information which is disseminated electronically to commanders and units after vetting by a staff officer who performs cognitive analysis. Physical reports may also be created and filed. The information terrain analysis can broadly be summarized as under subject to caveat of requiring further granularity:

Data Source	*Sensor and Acquisition*	*Data Generated*	*Transmission*	*Analytical Processing*	*Storage*
Natural and man-made objects	Electronic and cognitive	Electronic and physical	Mostly electronic (both radiating and non-radiating)	Electronic (hardware + software) and Cognitive (analysis by Staff Officers)	Electronic and Physical

This example is actually a case of multiple information cycles (a large piece of ground), lying predominantly in the electronic realm. For example it contains patrols (Example 2 above), Radars (Example 3 above) and a Networks (similar to Example 4 above).

The fact that the terrain being analysed can be of varying dimensions should not be surprising. The terrain analysis which a division commander does for his area of responsibility is obviously not the same as what one of his company commanders would do with respect to the specific landmark which it is his responsibility to attack or defend. The granularity keeps varying with the level.

The above examples tell us that it is indeed possible to carry out terrain analysis in the information domain and use it to formulate plans for offensive as well as defensive information operations. Attributes can be customized and the information cycle can be sub-divided into increasingly smaller pieces of terrain depending on the level at which the appreciation is being carried out.

Consider a motivational speech expected to be given by a leader before outbreak of hostilities and to be broadcast via radio and television. An IW unit commander tasked to disrupt (or monitor) such a broadcast must use a more detailed set of attributes while carrying out his terrain analysis. It may look as under:

- Data Source – Human (the leader)
- Sensor and acquisition – Electronic and acoustic (camera/mic)
- Data generated – Electronic (digital video/audio)
- Transmission – Electronic/Radiating (UHF/VHF/HF)
- Analytical Processing – Electronic (broadcasting system) and Cognitive (minds of the audience)
- Storage – Electronic (broadcasting system) and Cognitive (minds of the audience)

The IW unit commander may also like to break up the overall information cycle (terrain) into smaller cycles – one taking place in the broadcaster's studio, one during the broadcast itself and one at the individual audience's level. An option emerges at each level. The attacker can thus target the studio infrastructure, the broadcaster's internal network or the broadcasted frequency to monitor or prevent the audience from receiving the broadcast. The best approach (or combination of approaches) can now be discerned depending on the mission.

More detailed the information terrain analysis greater will be the number of options that will emerge. Also the analysis is not restricted to offensive operations. The same analysis done from the defender's point of view will yield the enemy's most likely approaches, the design of defense and the resources required to beat back the attack. The technique is very similar to what one does in other battlespaces. A well trained IW professional will be quickly able to do such an analysis and not only arrive at the best plan but also dynamically synchronize it with operations in other battlespaces. For example if kinetic attacks are required to be dovetailed into the IW plan, it will become obvious from the terrain analysis. Similarly feasibility and requirement of sabotage or espionage operations instantly becomes apparent.

It is also important to understand that just as a commander with a smaller force may choose to infiltrate rather than directly attack a strongly held position; similarly operations and tactics in the information battlespace are always and invariably linked to sufficiency of resources. If a jammer with requisite frequency range and power is not available then the option is not practically viable and another option must be exercised. Such alternative options can also lie outside of the information battlespace (for example use of kinetic means delivered by land, sea or air to disrupt the information cycle). At the strategic level such holistic analysis points to the diverse set of offensive and defensive capabilities that need to be created and force structures that will optimally harmonize their application.

Another very important lesson which emerges from the above discussion is that while electronic data and information is becoming increasingly predominant, the information cycle can however get completed entirely in the cognitive or physical domains bypassing the electronic realm. Some of the most valuable information in fact lies buried in the minds of people or locked away as documents in heavily guarded secure vaults. Electronic devices are often forbidden from sensitive meetings. Handwritten notes are not allowed at many classified briefings. It is also a policy in many organizations that Top Secret data (or information) is not stored or processed in electronic format. When leaders or diplomats meet or when businessmen arrive at a deal; the information cycle often bypasses both the physical and electronic format.

Clearly if information is to be treated as a contiguous battlespace, then we cannot exclude the abstract and physical forms of information simply because data has not been stored or transmitted electronically. Technology heavy IW doctrines are like stating that we will only fight in mountains but not in the deserts or jungle. However, integrating the entire operational capability of the three terrains under a single umbrella has its own set of challenges. We shall go into more details subsequently.

Peculiarities of the Information Battlespace

The last important aspect is that though we have used an analogy with land for ease of understanding; each battlespace has its own peculiarities. Information also has several of them and these are highlighted below.

- **Polymorphism.** Just like the element Carbon can exist as coal, diamond or graphite; information exhibits similar polymorphism and can concurrently exist as cognitive, physical and electronic. The same password for example can be memorized (cognitive), written on paper (physical) and stored on a server (electronic). Targets can also rapidly move from one type of terrain to another. A wireless transmission (or a phone call) exists very briefly in the electronic terrain before it transforms into a physical log or gets absorbed as cognitive information. Appearances can therefore be deceptive and care needs to be exercised in classifying information into a particular type. A CD containing secret electronic data but lying locked inside a vault is similar to a physical paper file in many ways. It is more vulnerable to operations tailored for attacking physical information then electronic. Several such situations exist where operations pertaining to a different type of terrain are more suitable depending on the disposition of the target.
- **Static and Mobile Targets.** Information under storage is a static target. These can be physical documents, electronic data or information within the human mind. However, information by nature is meant to be communicated and therefore a substantial portion of it is also moving from one point to another at any given point of time. This can be in the cognitive realm (words being spoken), physical realm (documents and objects being transported) or the electronic realm (transmissions). These are the moving targets. Static and mobile targets require different type of tactics and equipment.

- **Time Sensitivity.** Moving targets in conventional battlespaces (viz tanks, ships, missiles, orbiting satellites and aircrafts) have a finite velocity. This gives us time to detect them and maneuver resources to intercept them. Electronic information on the other hand can move at the speed of light. A spoken sentence similarly exists only for a brief moment. If such moving targets are to be engaged then operations must be executed in synchronicity and in real time. Akin to an ambush, one has to be ready and waiting at the right place and time else the opportunity is lost.
- **Non-Linear Battlespace.** War in the information battlespace is non-linear. The outcome of operations is not necessarily related to the resources employed and disproportionate results can be accomplished by committing a small force and at a low cost. Operations also do not follow a linear timeline and multiple targets can be engaged simultaneously irrespective of their disposition and physical location. Information being a virtual, contiguous and omnipresent battlespace; it allows attacks (viz those involving the Internet) to be launched from anywhere in the world and at any time thereby making it easier to surprise the adversary. Lastly, information operations can be conducted both during peace and war.
- **Fluidity.** The information landscape is fluid and evolves at a much more dynamic pace as compared to any other battlespace. New types of sub-terrains (collection of targets and objectives with similar attributes) may emerge and existing ones may quickly vanish. This is particularly true of the electronic terrain where the life span of technologies, systems and platforms may be as short as five to ten years. This puts tremendous stress on operational doctrines, equipment, planning and force structures which must keep pace.
- **Cognitive Inaccessibility.** The cognitive terrain which resides within the human mind is a black box which can only be accessed via the physical or electronic route. While the human mind constantly generates and analyzes information by itself, such information cannot be accessed unless it is communicated via speech, gestures, writing, electronic messages etc. Any attempts to attack cognitive information cycles must therefore rely on delivering an external stimulus in the form of physical or electronic information.

- **Continuity and Overlaps.** As seen in the information terrain analysis examples, objectives in the information battlespace often straddle several types of terrain at once. One way to visualize this paradox is to compare the information landscape with the electromagnetic spectrum. While we see only a mix of colours (ie white light), yet it can be split into individual components and also grouped into categories (red, blue, green or UV, visible, IR etc) if we want to. For example, Internet based social media straddles both the cognitive and electronic terrains. This continuity and overlap can be diagrammatically depicted as under:

Figure 2.4: The Overlapping Nature of Information Terrain

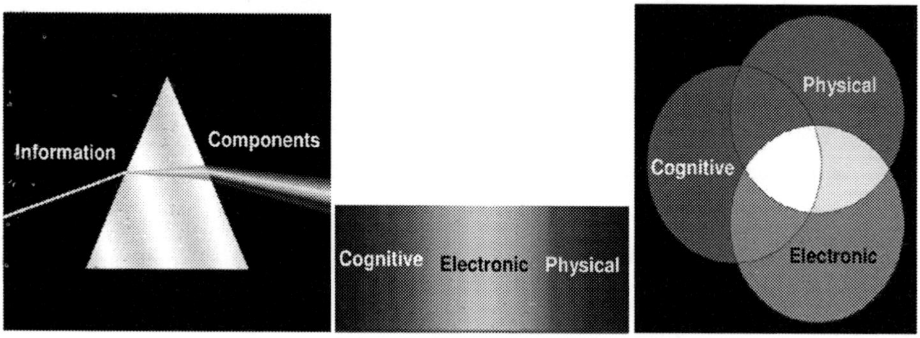

One reason for the continuity and overlap described above is the constantly expanding scope of electronic information as a result of which most targets in the cognitive and physical terrains are gravitating towards the electronic. This explains why IW is increasingly getting associated with the electronic realm of information.

- **Electronic Sub-Terrains.** Amongst the three basic types of terrain it is the electronic terrain which is now the most predominant and vast. This vastness is not only in terms of the volume of information but also its technical diversity. One may compare it to the vast Himalayan mountain range which has low lying hills, high mountains, extreme high altitudes, plateaus, cold deserts as well as glaciers. Precipitation and snowfall similarly varies a great deal and modulates the terrain. Some of the resultant sub-terrains are so peculiar that they call for unique type of operations, tactics and equipment. Similarly, two very distinct sub-types can be identified within the larger electronic terrain. These are described below:

- *Computational Devices.* The exponential increase in the quantum of electronic information requires that enormous amounts of computational resources (hardware and software), digital data storage and networking capability must be committed for information to be processed, retrieved, analysed and communicated. So vast and unique is this rapidly expanding group of electronic computational devices having hardware, software, digital data storage and networking capability that the underlying ecosystem may be deemed to be a special sub-terrain amenable to a distinct class of operations. It also requires to be mentioned that such electronic computational devices are no longer restricted to servers, desktops or laptops. It includes any device which has hardware, software, data storage and which is capable of electronic interaction. Thus smart phones, routers, intelligent sensors, smart televisions, smart cars, biometric devices all fall within this category. It is not relevant whether such a device is actually connected to a network or not. Like an aircraft on ground, a ship in port or a reserve battalion in depth awaiting orders to join combat, the potential exists and the target can be engaged depending on its disposition. This sub-terrain comprising of electronic computational devices is rapidly expanding and many targets in the physical and cognitive terrain are now moving towards it. Cyber warfare has emerged as a key operational capability in this sub-terrain.

- *Electronic Transmission.* With the increase in volume of electronic information; bulk of the information under transmission today comprises of data modulated onto electromagnetic waves and travelling over copper cables, optical fibers, and wirelessly through free space as well as underwater. So vast is the number of these moving electromagnetic energy targets that electromagnetic transmission requires to be deemed as another key type of sub-terrain within the electronic type. Within this sub-terrain, radiating (wireless) and non-radiating (copper cables, optical fibers) are two major sub-types which require distinct types of operations. Radiating transmission over wireless media in particular plays a vital role in enabling mobility in the battlefield. It is also much easier to acquire and attack such wireless transmissions vis-a-vis non-radiating media which require covert and sustained physical access to the wires and cables. EW which targets the wireless transmissions and SIGINT which targets both wireless and wired transmissions are two key operational capabilities within the transmission sub-terrain.

Summary of the Terrain Based Framework

The need for a theoretical framework to describe the information battlespace and the operations which can be conducted within it is because it helps us understand the logic why things must be done in a certain way, and what are the possible options.

Several other theoretical frameworks for IW exist and it would be appropriate to mention these for the sake of completeness. Some of these are as under:

- The OODA loop based model of IW
- The information and decision theory based model of IW
- The game theory based model of IW

These theories convey many key ideas which have played an important role in formulation and evolution of IW doctrines over the past decades. The reason for laying down a new terrain type theoretical framework is because it enables an information warrior (and there are all types) to quantify the objective and plan actual operations.

Finally it is time to summarize and describe the information battlespace in terms of our terrain based framework discussed above. It would look something as under:

Figure 2.5. Major Types of Terrain in the Information Battlespace

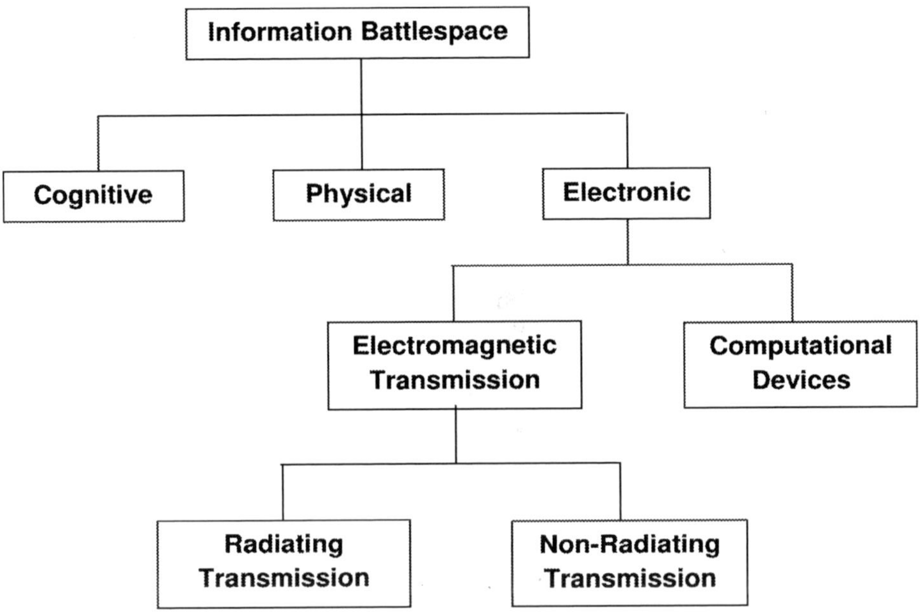

It deserves an answer as to why it has not been attempted to subdivide the cognitive and physical terrains on similar lines as the electronic, or why the electronic terrain has not been subdivided further. The reason is that we are only interested in major categories which lead to a distinct class of operations and where a large number of targets congregate. More sub-categories do not necessarily add much value to our analysis of IW force structures. The framework is open ended and can certainly be expanded beyond if needed. A unit commander conducting actual information operations must certainly do so if required.

Having dealt with the terrain, it is now time to explore the operations, tactics, technologies, platforms and weapon systems that manifest in this new battlespace.

ENDNOTES

1. Martin Hilbert and Priscila López, "*The World's Technological Capacity to Store, Communicate, and Compute Information*", (Science magazine, Vol. 332, Issue 6025, April 2011), pp.60-65.

SECTION 2

Conduct of IW

CHAPTER 3

Operations and Capabilities

Having seen that the information battlespace comprises of three main types of terrain namely electronic, physical and cognitive; the next logical step is to inquire as to what type of operations are possible in these terrains and what capabilities are required to fulfill missions at the strategic, operational and tactical levels. We again start with first principles and use the terrain based framework to guide us further.

A time tested approach is to identify the basic operations of war and then superimpose them on the terrain. For example in case of ground warfare, the two most fundamental operations are attack and defense. We can then combine these basic operations with various types of terrain to arrive at an operational framework. Such a framework will consist of operations like attack in mountains, defense in mountains, attack in deserts and defense in deserts etc. The layout of defenses, the methods of attack, the tactics and targets automatically become apparent once we have such a framework. Forces can be grouped and equipment can be planned accordingly. Also once operational capability is grouped in such logical manner; it can then be applied across a range of objectives and targets based on the overall design of battle.

Let us therefore first try to identify what are the basic operations of war possible in the information battlespace.

Major Operations of War in the Information Battlespace

Three basic types of operations can be identified in the information battlespace. These are as under:

- **Acquisition.** Acquisition is a fundamental operation in the information battlespace. It helps to discern the enemy's capability, disposition, intent and plans. It also generates targets for attack and can identify own vulnerabilities which need to be defended. Acquisition is also called exploitation. Acquisition operations comprise of dozens of niche minor operations depending on the manner in which information is acquired in the cognitive, physical and electronic terrains.

- **Attack.** Attack is the second basic operation in the information battlespace. Attacks are directed against constituents of the information cycle ie sensors, data acquisition, transmission, analytical computing and storage or the resultant data and information itself. Attacks require prior acquisition operations to identify targets and vulnerabilities. Several types of attack are possible and broadly speaking they fall under the following categories:
  - *Degradation and Manipulation.* These attacks are designed to diminish the quality of data and/or information. Manipulation (or deception) is a special type of degradation where customized and genuine looking misinformation is injected into enemy's information cycle with an aim to produce a tailored response.
  - *Disruption and Denial.* These attacks are designed to diminish the volume and velocity of data and/or information. When the disruption is prolonged, it may be termed as denial.
  - *Destruction.* These attacks are designed to permanently destroy data and/or information. It differs from disruption and denial because of its permanent and irreversible nature.

  Attacks against information can also be scripted in other battlespaces (viz kinetic attacks to degrade, disrupt, deny or destroy the information cycle constituents).
- **Defense.** Defense is the third basic operation of war in the information battlespace. Defensive operations involve early detection of threats, thwarting acquisition attempts and protecting the information cycle constituents against attacks.

Combining the three basic operations of war with the major terrain classifications discussed earlier gives us the following set of basic major operations that maybe conducted in the information battlespace:

Terrain	*Basic Operation*	*Type of Major Information Operations*
Cognitive Information	Acquisition	Acquisition of cognitive information
	Attack	Attack on cognitive information
	Defense	Defense of cognitive information
Physical Information	Acquisition	Acquisition of physical information
	Attack	Attack on physical information
	Defense	Defense of physical information

Terrain	*Basic Operation*	*Type of Major Information Operations*
Electronic Information	Acquisition	Acquisition of electronic transmission
		Acquisition of computational information
	Attack	Attack on electronic transmission
		Attack on computational devices
	Defense	Defense of electronic transmission
		Defense of computational devices

It needs to be remembered that attacks on cognitive information cannot be carried out directly. Instead the attack relies either on preventing sensor acquisition or on injection of malicious cognitive, physical or electronic information at the data source and sensor acquisition stage.

It also needs to be recalled that attack and defense in the physical realm of information require physical operations to be carried out.

Operations in Detail

All major operations of war comprise of numerous minor operations. For example infiltration, heliborne/airborne/amphibious operations, patrolling, raids, ambushes etc are distinct minor operations conducted as part of the major ground operations. Similar is the case in the information battlespace with the only difference being that the vastness and diversity of the information battlespace results in a much longer list of such minor operations.

Also as already brought out, the idea of treating information as a unified battlespace and consolidating all information related operations under the overarching umbrella of IW is a recent phenomenon. Else many of the underlying disciplines already existed independently and have their own associated terminology which finds mention in existing doctrines and literature. It is therefore also important to try and reconcile the terrain based operational framework derived above with already existing IW terminology and jargon. A classification based on targets, equipment, tactics and missions is presented below. The list is not exhaustive but sufficient to illustrate that the terrain based operational framework can be easily linked to existing IW terminology.

- **Target or Equipment Specific Minor Operations**. These minor information operations are directed against specific targets (or threats in case of defense) within a particular type (or subtype) of information terrain. They may also be based on a specific type of equipment. Some examples are given below:

Type of Operation	*Associated Terminology*
Manipulation attack on cognitive information	Propaganda, Perception Management, Media Warfare, Public Affairs, Strategic Communication, Fake News, Social Media Engineering
Acquisition of physical information through satellite imagery and aerial photography	Imagery Intelligence (IMINT)
Acquisition of electronic information (communications only)	Communications Intelligence (COMINT)
Acquisition of electronic information (non-communication only)	Electronic Intelligence (ELINT)
Acquisition of encrypted information by breaking the cipher algorithms	Cryptanalysis
Denial of satellite navigation	Navigation Warfare (NAVWAR)

Observe how these involve only a very specific type of operation (viz manipulation attack only) or a specific type of equipment (satellite imagery) or engage a very specific target (viz encryption algorithms only). Many niche target and equipment specific information operations like Technical Intelligence (TECHINT), Measurement and Signature Intelligence (MASINT), Financial Intelligence (FININT) have come up as the information battlespace expands.

- **Tactical Synergy Based Minor Operations.** These minor information operations are based on commonality of tactics. Some examples are as under:

Type of Operations	*Associated Terminology*
Acquisition of cognitive, physical and electronic information (through human effort only)	Human Intelligence (HUMINT)
Acquisition of cognitive, physical and electronic information (from publicly available sources only)	Open Source Intelligence (OSINT)
Defense of physical and electronic information from acquisition (by encryption)	Cryptology
Acquisition of electronic information under transmission (radiating only)	Electronic Support Measures (ESM), Electronic Support (ES)
Acquisition of electronic information resident inside computational devices	Computer Network Exploitation (CNE)

Type of Operations	*Associated Terminology*
Transmission attack (radiating only)	Jamming, Electronic Counter Measures (ECM), Electronic Attack (EA)
Attacks on Computational Devices	Computer Network Attack (CNA)
Defense of electronic information under transmission (radiating only)	Electronic Counter-Counter Measures (ECCM), Electronic Protection (EP), Electronic Emission Security (EES or EMSEC)
Defense of electronic information against computational attacks	Cyber Security, Computer Network Defense (CND), Defensive Cyberspace Operations (DCO)

Observe how HUMINT (acquisition of information by spies and agents), OSINT (acquisition of information from public sources) and cryptology (encryption of electronic and physical messages) cut across terrain types and targets but share common tactics. Likewise amongst transmission, only the radiating type sub-terrain has been singled out in case of ESM, ECM and ECCM because operations involving non-radiating transmissions require different tactics and equipment.

- **Mission Specific Operations.** It is also possible to have information operations orchestrated to achieve a well-defined mission. This is particularly true of acquisition operations wherein it is desirable to acquire as much information as possible for ease of correlation and corroboration regardless of the type of information terrain or sub-terrain. Examples are as under:

Type of Operations	*Associated Terminology*
Acquisition of all electronic information (radiating transmission, non-radiating transmission as well as information resident inside computational devices)	Signals Intelligence (SIGINT)
Defense of all electronic transmissions (radiating and non-radiating) against unauthorized acquisition	Communications Security (COMSEC)
All information acquisition operations	Military Intelligence, Intelligence, Surveillance and Reconnaissance (ISR)
All types of deception in cognitive, physical and electronic realm	Military Deception (MILDEC)

In the above examples the tactics, terrain as well as targets are diverse. The operations may even require synchronization across multiple battlespaces (viz ISR). Yet combining these diverse operations into a single capability contributes in a major way to the war effort thus justifying such mission based grouping of information operations.

- **Inter-battlespace Minor Operations.** Inter-battlespace minor operations are those where the execution is not confined to the information battlespace yet these are essential for acquisition, attack or defense. Examples are:

Type of Operations	*Associated Name(s)*
Acquisition of cognitive, physical and electronic information using spies	Espionage
Attacks on physical and electronic Information targets using infiltrators, agents or third columnists	Sabotage
Kinetic attacks on physical and electronic Information targets	Physical destruction
Defense of physical and electronic information targets against espionage and sabotage	Physical security

These inter-battlespace operational capabilities may or may not be deemed to be information operations, however, discarding them from the purview of IW leads to major loss of warfighting potential in the information battlespace. One viable option may be to treat these as supporting capabilities.

Capabilities

As seen above, the twelve basic type of major information operations result in many different minor operations depending on the terrain, tactics, type of target and mission. Since it is impossible to create independent force structures for each one of them, we attempt to group them into synergistic capabilities. The word "capability" denotes a potential (or ability) to carry out a specific task or accomplish a specific mission or assignment. Defining capabilities is essential to formulation of operational doctrine and creation of force structures.

IW being an evolving idea, there is no standard set of capabilities. At the highest level IW itself may be deemed as a capability or else we may pick up a very niche type of operation like disrupting satellite based navigation systems (NAVWAR) and call it as a capability. The reason for defining a type (or grouping) of operations as a "capability" lies in the belief that such capability can contribute towards achieving war objectives in a major way. Once a capability is identified based on this criteria then organizations, equipment, training, financial resources and manpower can be suitably dedicated to operationalise it. Operational doctrine and force structures are concurrently evolved to ensure proper coordination of effort and tasking of resources.

The 2006 version of the US IW doctrine JP3-13 for example identified EW, CNO, Psy Ops, MILDEC and Operations Security (OPSEC) as "core capabilities" whereas Information Assurance, Physical Security, Physical Attack, Counterintelligence and Combat Camera were deemed to be "supporting capabilities". Civil-military operations, public affairs and public diplomacy were deemed to be "information related capabilities"[1].

The 2012 revised version of the doctrine did away with the classification as core, related and supporting capabilities and instead designated all of them as Information Related Capabilities (IRCs). It renamed Psy Ops as Military Information Support Operations (MISO). It also replaced CNO and its sub-categories of CNE, CNA and CND with Cyberspace Operations (CO) having three new sub-categories of Department of Defense Information Network (DoDIN), Defensive Cyberspace Operations (DCO) and Offensive Cyberspace Operations (OCO)[2]. The revised doctrine lists fourteen Information Related Capabilities (IRCs), namely Joint Electromagnetic Spectrum Operations (JEMSO), Cyberspace Operations (CO), Military Information Support Operations (MISO), Military Deception (MILDEC), Operations Security (OPSEC), Information Assurance (IA), Intelligence (Intel), Strategic Communication (SC), Public Affairs (PA), Special Technical Operations (STO), Space Operations (Space Ops), Civil Military Operations (CMO), Key Leader Engagement (KLE) and Joint Interagency Coordination Group (JIACG)[3].

The US doctrine defines IRCs as "*Tools, techniques, or activities using data, information, or knowledge to create effects and operationally desirable conditions within the physical, informational, and cognitive dimensions of the information environment*"[4]. In terms of our terrain based framework "capabilities" are a synergistic grouping of operations which fits into the overall warfighting doctrine and contributes significantly towards achieving war objectives.

The point to be understood is that we may define as many capabilities as we want subject to the basic criteria that they contribute towards achieving war objectives in a major way and are supported by operational force structures. No two countries (other than US allies) follow the same definition nor have other countries defined the capabilities in a similar fashion. Even the US/NATO doctrine is constantly evolving.

Core Capabilities

Despite there being no standard list of capabilities, four of them stand out. These are Psy Warfare, Intelligence, EW and Cyber Warfare and the reason they stand out is because collectively they cover almost the entire information terrain as well as the underlying operations. All other capabilities are either a subset of these or share a close synergy and interdependency. For example the US defined IRCs of Strategic Communication, Public Affairs, Civil-Military Operations and Key Leader Engagement are mostly variations or subsets of Psy Warfare without stating so in obvious terms. SIGINT and HUMINT are similarly part of the intelligence capability.

Another reason for calling them as core capabilities is because all modern armed forces have extensive force structures (ie headquarters, units and equipment) dedicated to these four capabilities. This is unlike MILDEC which is mostly a staff planning function or OPSEC which is predominantly a process or method. There is no dedicated force structure associated with either of these.

As brought out while discussing peculiarities of the information battlespace, targets tend to straddle multiple terrains as also exhibit polymorphism. This results in capabilities also being forced to straddle multiple terrains in order to exert their full potential. The core capabilities of Psy Warfare, Intelligence, EW and Cyber Warfare therefore have one associated primary terrain, but also exert influence in other types. These can be deemed as secondary (or tertiary) areas of influence. The overlaps lead to debate but there is no way out of this conundrum.

It also merits attention that while Psy Warfare, EW and Cyber Warfare combine all the three basic operations of acquisition, attack and defense; Intelligence is thought of as being restricted to acquisition only. This understanding is incorrect because intelligence operations involve offensive actions like sabotage as well as defensive actions like counter-intelligence. Offensive and defensive capabilities in the physical terrain are as vital as those in the cognitive and electronic types. Intelligence ops cover this space.

The table below tries to summarize these aspects along with some explanatory remarks:

Capability	*Primary Operational Terrain*	*Secondary Operational Terrains*	*Remarks*
Psy Warfare	Cognitive	Physical and Electronic	Psy warfare operations are cognitive; but cognitive perceptions can only be accessed through physical and electronic means.
Intelligence	Physical	Electronic and Cognitive	Use of spies and agents to gather information and carry out espionage or sabotage in the physical domain are the primary intelligence functions. Yet these soldiers must also operate in the electronic and cognitive terrains in order to produce results. Corroboration of intelligence likewise requires operations to straddle the cognitive as well as electronic terrains. SIGINT for example is a major component of intelligence straddling the electronic terrain and the only capability which targets the non-radiating transmissions.
EW	Electronic (Radiating Transmission)	Cognitive and Computational Devices	EW primarily targets radiating transmissions but can also be used as a vector to deliver psy warfare (cognitive) audio-visual broadcasts. Since many modern computational devices (like smart phones) are also transceivers, EW can also facilitate acquisition of cyber warfare targets and delivery of malicious payloads.
Cyber Warfare	Electronic (Computational Devices)	Cognitive and Electronic (Radiating and Non-Radiating Transmission)	Cyber warfare primarily targets computational devices but the increasing role of Internet based platforms in shaping perceptions implies that it also has a footprint in the cognitive terrain. Cyber warfare also has an increasing role to play in operations involving transmission. This is due to the convergence of communication and computing wherein most of the switching and processing is now handled by computational devices.

Let us look at these core capabilities in slightly greater detail.

Psy Warfare

Psychological Warfare (Psy Warfare) is the oldest amongst the various IW capabilities. Psy warfare is aimed at influencing the target audience's value system, beliefs, emotions, motives, reasoning, or behavior to induce attitudes or decisions favorable to the attacker's objectives. It is also used to destroy morale of the enemy as well as to gain support of neutral audiences. In terms of our terrain based model, psy warfare lies in the cognitive terrain.

Psy warfare is conducted both during war and peace and by state as well as non-state actors (viz terrorist organizations). Individuals and organizations like political parties, social media activist, corporate and business houses also conduct psy warfare within permissible legal limits. Political warfare, propaganda, advertising, Winning Hearts and Minds (WHAM) are all variants of psy warfare. Targets are not limited to soldiers or military commanders and often include political leadership, government agencies, non-governmental organizations, religious groups and communities.

Since the human mind can only be accessed through external stimulus, psy ops must rely on electronic and physical realms to achieve their objective. Electronic mass media and Internet have become the preferred vectors for delivery of such psy ops payloads in the information age. Vectors in the physical terrain (direct communication, newspapers, posters and pamphlets) still remain extremely relevant, particularly in the warzone. We may compare this situation to amphibious or airborne operations where the approach to the objective lies through sea/air but the objective is on land. Terrain encountered during the approach as well as the one at the objective modulate the conduct of operations.

Amongst all the IW capabilities psy warfare is the most complex and difficult to quantify yet it yields the greatest dividend because human decisions are ultimately made in the cognitive realm. If decision making is relegated to computational devices (artificial intelligence) in the future, we may see a convergence of psy and cyber warfare. That however is not the present scenario and psy warfare remains a distinct and decisive capability. We will restrict ourselves to the military aspects.

Psy warfare maybe waged at strategic, operational or tactical level and often concurrently. The distinction is explained below:

- **Strategic Psy Ops**. Strategic psy ops are conducted at the national level both during peace and war. These are closely supported by diplomacy as well as other organs of the government. Civilian organizations may also be co-opted for greater effectiveness. The objectives are usually long term in nature with the aim often being to either avoid armed conflict or enter conflict in an environment that is favorable. Strategic psy warfare is also aimed at gaining cooperation from friendly and neutral nations while preventing similar assistance being received by the adversary.
- **Operational Psy Ops**. Psy ops at the operational level are conducted by the armed forces in order to support military objectives. Targets include enemy combatants, insurgents, non-state actors, terrorist organizations as well as the local population. Operations are generally conducted in distinct geographies or theatres and may selectively include local civil administration and organizations on an as required basis.
- **Tactical Psy Ops**. These are local psy ops carried out by tactical commanders in support of their mission. They are often in furtherance of psy ops planned at the operational level but may also be conducted in isolation to break the enemy's morale and will to fight.

Acquisition operations in case of psy warfare are aimed at acquiring targets for attack and monitoring the impact (damage assessment). Intelligence operations, particularly HUMINT, espionage and SIGINT play a major role in acquiring such inputs. No meaningful offensive or defensive psy ops can be launched without acquisition operations.

Counter psy ops is the defensive component of psy warfare aimed at protecting own forces and friendly elements from enemy's cognitive attacks. Pre-empting and countering disinformation spread by the adversary and dissemination of own intentions and activities are the key activities undertaken as part of counter psy ops.

Attack operations vary a great deal depending on the disposition of the target and they have evolved over time. Historical methods of waging psy warfare included spreading of ruses or rumor, direct communication in the form of speeches and discussions, use of loudspeakers, posters, pamphlets, radio broadcasts, newspapers, books as well as use of songs, films and documentaries. These were joined by television which became a key vector for disseminating psy warfare products due to its increasing reach and popularity. Internet and electronic mass media have now emerged as the preferred vector for delivery of psy op products. Unlike conventional psy ops, the ICT version requires continuous engagement with the target audience, monitoring of responses and modifications or refinement to the product and/or theme in near real time. Threat of legal action or economic sanctions is also often used as a form of psy warfare. Nation states may also use military posturing as part of psy warfare. Threat of physical or material harm is likewise used to wage psy warfare especially by non-state actors.

The process of conducting psy ops is as explained in the flow chart below:

Figure 3.1: Conduct of Psy Ops

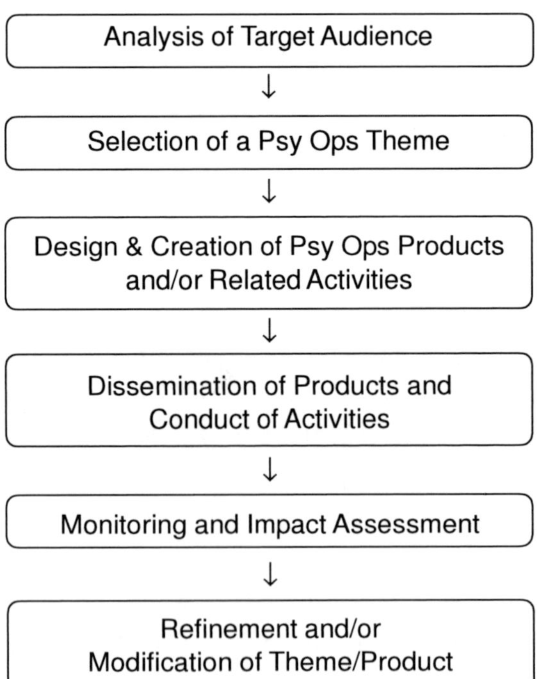

Intelligence

Intelligence is another core IW capability and again much older than the concept of IW itself. In the early days intelligence operations were primarily restricted to acquisition of information using spies and agents i.e. HUMINT. It also included espionage and sabotage operations. In terms of our terrain based model, intelligence thus covered the entire physical information terrain and also touched the cognitive. The scope of intelligence operations expanded once electronic information became predominant. It is only through acquisition operations cutting across the entire information battlespace that intelligence can be corroborated and refined. While acquisition remains the main operation, intelligence also includes attack (sabotage, espionage) and defense (counterintelligence) operations.

Being the only capability which manifests in the physical terrain, the other core capabilities of psy ops, EW and cyber warfare must invariably integrate with intelligence operations in order to reach their full potential. Conduct of insider attacks on computer networks, tapping of fiber optic cables through sabotage and gauging impact of psy ops through HUMINT are examples where electronic and cognitive operations are facilitated by intelligence operatives present on ground.

Today intelligence comprises of a vast array of special operations and disciplines. Some of these are enumerated below:

- **HUMINT**. HUMINT is a major sub-class of intelligence operations involving acquisition of information from persons on ground and a key sub-capability. HUMINT may acquire both cognitive as well as physical information (documents, photographs etc) and has a vast range of targets. HUMINT operations are conducted by both military and civilian intelligence agencies and they may also collaborate in many cases depending on the nature and disposition of the target. Military HUMINT operations are conducted by intelligence units and special reconnaissance teams. Patrols and military police personnel also provide HUMINT. Interrogation of prisoners of wars, captured terrorists, criminals, enemy agents and refugees is another valuable source of HUMINT. Diplomatic staff, Non-Governmental Organizations and debriefing of friendly personnel (viz tourists) can also provide HUMINT. Psy warfare capability relies on HUMINT to identify targets and ascertain efficacy of operations.

- **Espionage.** Espionage is yet another key intelligence sub-capability which involves gaining covert information either through spies or by recruiting insiders and local sources. Espionage operations are closely related to HUMINT but have now diversified into the electronic realm. Industrial espionage through cyber warfare is an example.
- **SIGINT.** SIGINT is yet another major class of intelligence operations and a distinct sub-capability. The origins of SIGINT can be traced back to the invention of wireless communications and it matured during World War-I. Agencies dedicated to encoding and decoding messages also came up. Cryptanalysis became a key subcomponent of SIGINT during World War-II. Breaking of the Enigma cipher used by Germany was one of the decisive factors behind the victory of the allied forces and numerous computational devices were developed and deployed to break cipher codes. SIGINT diversified into COMINT and ELINT post the invention of radars. Over the years it has become a dominant intelligence gathering capability targeting every type of transmission whether radiating or non-radiating. Thus tapping into submarine optical fiber cables, telecommunication networks and Internet gateways is also part of SIGINT. Operations often utilize organizational synergy with other intelligence disciplines like espionage and sabotage. Cryptology has remained an essential sub-component of SIGINT with both offensive (code-breaking) and defensive (code-making) operations. All modern SIGINT organizations have enormous computing resources at their disposal to facilitate code breaking and also analyze the huge volumes of electronic data that they intercept. Big data analytics and deep packet inspection are thus part of SIGINT operations. In terms of our terrain based model, SIGINT spans acquisition operations across the entire electronic transmission realm. Since acquisition operations are also fundamental to both attack and defense; SIGINT shares synergy with the other two electronic terrain capabilities of EW and cyber warfare. Cyber warfare in particular has become inextricably linked to SIGINT due to the convergence of computers and communications. The overlaps pose a dilemma when it comes to consolidating the IW capabilities.
- **Geospatial Intelligence (GEOINT).** GEOINT covers intelligence gathered from satellite imagery, aerial photography and other sources of geospatial information. Imagery Intelligence (IMINT) is a subset.

- **Measurement and Signature Intelligence (MASINT).** MASINT is a specialized discipline involving intelligence derived from data captured using sensors. It has a large number of special operations covering acquisition, measurement and analysis of a diverse set of signatures acquired through electromagnetic, electro-optic, seismic, acoustic, nuclear, biological and chemical sensors. MASINT forms a key component of ballistic missile defense. MASINT and SIGINT share an overlap.
- **Imagery Intelligence (IMINT).** IMINT involves collection of physical intelligence using satellite and aerial photography. Geospatial Intelligence (GEOINT) is a wider term which includes data obtained from maps and may also superimpose demographic information.
- **Open Source Intelligence (OSINT).** OSINT involves intelligence acquired from publicly accessible sources to include Internet, books and publications, mass media, seminars and conferences etc.
- **Financial Intelligence (FININT).** FININT involves gathering intelligence from analysis of financial transactions. It plays an increasingly important role due to the interconnected and integrated nature of modern global economy. Much of it is now in the electronic terrain including crypto currencies but it also straddles the physical realm viz trust based informal systems like hawala which facilitate money laundering and terrorism.
- **Sabotage.** Sabotage operations are covert actions to subvert, disrupt or deny vital resource to the adversary. This also includes information. Sabotage is not a pure intelligence capability. Operations can be conducted by intelligence agents and also by infiltrating forces and third columnists. Targets maybe information or physical assets like railway lines. Insider attacks on computer networks, tapping of transmission cables, capturing enemy crypto material or devices are some examples where sabotage can facilitate information operations.
- **Counterintelligence.** Counterintelligence operations are defensive operations to thwart attacks on own intelligence organizations and assets by the enemy's intelligence agencies.

Intelligence is a vital and growing capability essential for planning operations in all battlespaces. All branches of the armed forces therefore have extensive intelligence force structures. It also facilitates all the other core IW capabilities.

Electronic Warfare

Electronic Warfare (EW), also called Electromagnetic Spectrum Operations (EMSO) is another core IW capability which pre-dates the IW construct. In terms of our first principle based terrain model; EW is a capability lying in the electronic transmission sub-terrain. EW however does not cover non-radiating transmissions which take place over copper/optical cables. The reason for this is because acquisition, attack and defense operations involving wired transmissions require substantially different tactics (viz sabotage), and equipment, vis-à-vis wireless transmissions taking place through free space.

Origins of EW can be traced back to the World War era when SIGINT emerged as a distinct capability to carry out acquisition operations against wireless transmissions. The acquisition operations were however soon joined by equipment and techniques to carry out attack (jamming, electronic deception) as well as defense (counter acquisition, counter jamming) and they were all synergistic and interdependent. Post World War-II the acquisition, attack and defense operations (and underlying equipment) therefore logically got consolidated into a singular distinct capability called EW and led to evolution of what we now refer to as "Integrated EW Systems" or "EW Suites" whereas SIGINT continued to exist as a capability restricted only to acquisition. Also, the ability of jammers to degrade enemy's command, control and surveillance capability led to EW becoming a part of Command and Control Warfare (C2W) and treated as a form of fire (like artillery) instead of being a part of intelligence.

Just like SIGINT branched off into COMINT and ELINT, similarly EW too was deemed to comprise of two related but technically unique disciplines of Communication and Non-Communication EW. This distinction has now become blurred. Today it is all about specific applications both in case of SIGINT and EW.

Both SIGINT and EW today exist as distinct and major capabilities even though the acquisition operations share major commonality. This often confounds some military thinkers who question as to why SIGINT can't be merged with EW and why must two separate hierarchies and organizations exist to conduct similar tasks. This belief is further reinforced when EW and SIGINT systems acquire identical information during war and counter-insurgency operations. It would therefore be appropriate to highlight some key differences between SIGINT and EW.

Firstly, as brought out above EW combines the acquisition operations with attack and defense. EW requires to conduct acquisition operations (search, monitoring and location fixing of emitters) in order to acquire targets for jamming and deception or for protecting weapon platforms against incoming attacks. EW suites mounted on aircrafts and ships are an example. Acquiring information is thus a facilitator as far as EW is concerned whereas it is the raison d'être in case of SIGINT.

Secondly, SIGINT has significantly grown from the World War days. It now also targets non-radiating communications and switching which are not covered by EW and which can only be accessed through sabotage which is part of the intelligence capability.

Thirdly, EW equipment is designed to engage radiating transmissions in the Tactical Battle Area (TBA) and EW systems are therefore mounted on mobile platforms and operate in close proximity to the enemy. They identify targets for electronic attack, protect platforms against incoming projectiles as well as provide real time situational awareness to commanders. Operations are conducted in real time and are dynamic. EW systems are thus designed and built keeping in mind the enemy's tactical systems. SIGINT systems on the other hand are designed to engage both military and non-military targets in a non-tactical setting. They are typically positioned at vantage points or in proximity to communications hubs so that they can acquire maximum intelligence.

Fourthly, cryptologic operations continue to remain part of SIGINT. The underlying infrastructure and organizations are closely integrated. Cryptologic analysis requires special resources and cannot be conducted in the TBA by EW.

Lastly, EW systems are closely integrated with kinetic means as also situational awareness and tactical decision support systems. This integration is achieved through hardware as well as software. In case of platforms like aircrafts and ships, EW equipment is also integrated with onboard sensors and includes passive countermeasures like chaff and decoys. All this leads to EW equipment being very different from SIGINT equipment in terms of design and architecture.

To summarize, the focus of EW acquisition operations is to generate targets for attack (jamming and deception) or defense (stealth and self-protection) as well as to provide situational awareness and integrate with kinetic weapons at the tactical level. SIGINT on the other hand operates at the strategic/ operational level and also covers non-radiating electronic transmission as well as cryptologic disciplines. There is very little commonality between EW and SIGINT in the modern time.

Coming back to EW – the scope of its operations has grown exponentially over the years as more and more entities in the battlefield rely on use of electromagnetic spectrum for command and control, situational awareness, synchronization and targeting. The spectral coverage has likewise expanded to cover the entire range starting from Very Low Frequencies (VLF) to Infrared (IR). The list of targets has concurrently expanded to include not just Combat Net Radio (CNR) and radars but also submarine communications, microwave links, satellites, cellular communications, navigational signals, ground to air communications, Radio Controlled Improvised Explosive Devices (RCIEDs), UAVs/UCAVs, missile homing/guidance and many other niche applications that use electromagnetic spectrum in the battlefield.

Modern EW is a highly specialized discipline with the latest development being the introduction of Directed Energy Weapons (DEWs). Specialized air, ground and seaborne EW assets are an integral part of all modern armed forces and sophisticated EW suites are standard fitment in all fighter aircrafts and battleships. Accordingly, multiple sub-specialties exit within EW. NAVWAR for example is a niche sub-capability which targets the satellite based navigation systems. Use of EW for anti-satellite (ASAT) operations is yet another fast growing sub-specialty.

EW technology is one of the most closely guarded secrets due to its decisive impact on the battlefield cutting across all types of battlespaces. One of the most profound examples of the impact of EW in recent times has been its employment by the US to suppress Iraqi command, control and air defense communications during the first Gulf War in 1991. The scope of EW has grown far beyond since then and today it is a major asymmetric capability which can also neutralize network centric advantages in the TBA.

In terms of application, EW is deemed to be a type of fire similar to kinetic weapons with the difference being that targets reside in the informational domain and energy delivered onto the target is electromagnetic radiation instead of kinetic energy. Deployment and employment principles of EW systems are quiet similar to kinetic-weapons.

Sub Components and Terminology

EW is subdivided into three major sub components of Electronic Support Measures (ESM), Electronic Counter Measures (ECM) and Electronic Counter Counter Measures (ECCM). These are explained below:

- **Electronic Support Measures (ESM) (NATO – Electronic Support (ES))**. ESM is defined as actions taken to detect, locate and classify sources of electromagnetic energy for purpose of threat recognition, targeting and/or planning conduct of future operations. In simple terms it is a consolidation of the acquisition operations in the wireless electronic transmission sub-terrain. Some important aspects to be remembered are as follows:
    - ESM is broadly divided into Search, Monitoring and Location Fixing (LF). The process is a continuous one and the main aim is to identify targets for attack. It is exactly similar to surveillance and reconnaissance operations carried out in land, sea or air battlespaces. ESM is also used to monitor success of attacks.
    - ESM activities may also be conducted to monitor own emitters for violation of Electronic Emission Policy (EEP), encryption policies, security classification, emission protocols or allotted frequencies i.e. it also supports defensive operations.
    - ESM also generates tactical SIGINT (Tac SIGINT) and situational awareness by discerning the location of enemy's headquarters, installations and movement of troops through location fixing of emitters or by listening into wireless transmissions taking place on the combat radio nets (which is increasingly rare because of the use of encryption). ESM thus helps generate the enemy's electronic order of battle (E-ORBAT) which can then be corroborated with inputs from other IW elements operating in the physical and cognitive realm to ascertain the enemy's intent and plans.
- **Electronic Counter Measures (ECM) (NATO – Electronic Attack (EA))**. ECM is defined as deliberate use of electromagnetic energy to degrade enemy combat potential or to deceive. It is the attack operation of war in the wireless electronic transmission sub-terrain. Some important aspects to be remembered are as follows:

- The targets are generated by ESM activity.
- ECM is considered as a form of fire.
- Use of chaff, flares and decoys are regarded as ECM because they deceive radars and homing missiles by degrading, manipulating and disrupting information about the target (viz a ship or aircraft). Anti-radiation missiles maybe considered a form of ECM though it is more appropriate to categorize them as a kinetic hard kill weapon and a form of physical destruction.
- **Directed Energy Weapons (DEWs)**. DEWs are an emerging ECM sub-capability and comprises of lasers, high power microwave and electromagnetic bombs. DEWs cause physical damage to the target using electromagnetic radiation as an alternative to kinetic weapons. The major difference is that kinetic projectiles travel with a finite velocity whereas electromagnetic energy (viz a LASER) travels at the speed of light and can hit a target almost instantaneously. The main challenges in development of DEWs are atmospheric attenuation, target tracking, ability to deliver sufficient power, miniaturization and mobility. Once these technological issues are overcome, DEWs will result in a paradigm shift in operational doctrine and tactics.

• **Electronic Counter Counter Measures (ECCM) (NATO – Electronic Protection (EP))**. ECCM is defined as measures undertaken to overcome hostile ESM or ECM. It is essentially the defensive operation in the wireless electronic transmission sub-terrain. Some important aspects to be remembered are as follows:

  - ECCM operations are conducted by all elements which rely on radiating electronic transmissions.
  - Operations include anti-jamming and anti-deception
  - Electronic Emission Security (EES) measures are part of ECCM.
  - Technical measures include designing receiver and antenna system that are resistant to ESM and ECM (viz by incorporating frequency agility, stealth techniques, frequency diversity, gain control, electromagnetic pulse (EMP) hardening etc)
  - Laying down the Electronic Emission Policy (EEP) to minimize the impact of enemy ESM and ECM is also part of ECCM.

The ESM (acquisition), ECM (attack) and ECCM (defense) operations can be depicted in a tree form as under:

Figure 3.2: The EW Tree

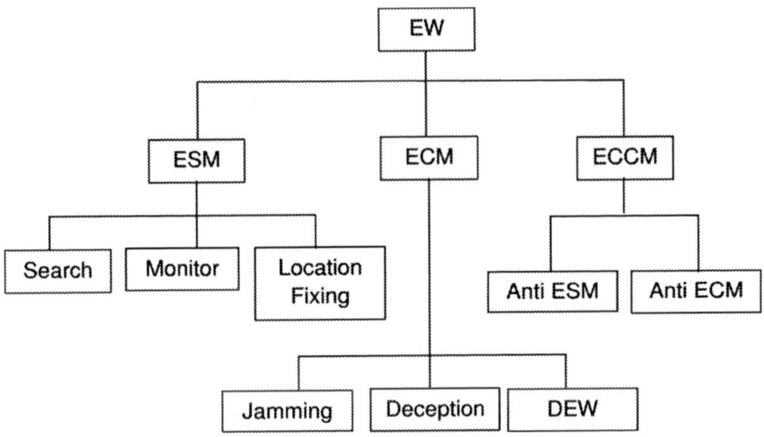

Cyber Warfare

In the terrain based framework which we have defined, cyber warfare comprises of a major class of operations in the electronic sub-terrain consisting of computational devices. It is the only capability which does not predate IW. In fact the whole idea of IW coincides with the emergence of cyber warfare as a distinct capability. Basic operations again comprise of acquisition, attack and defense. Distinction between civil and military targets is extremely blurred when it comes to cyber warfare. It is the most talked about capability, yet the most confounding because of the multiple terms and definitions associated with it as also its increasing overlap with other capabilities.

The terminology related confusion is mostly because of the changing US cyber warfare doctrine which has been widely quoted and copied. Till around 2012-13 the US doctrine used the terminology Computer Network Operations (CNO) as the collective designation of operations conducted as part of cyber warfare. CNO in turn was sub-divided into Computer Network Exploitation (CNE), Computer Network Attack (CNA) and Computer Network Defense (CND). This was fairly logical in that it broadly mapped onto the three basic operations of acquisition, attack and defense except for use of the term "network". The term "network" excluded computational devices which were not connected to the network and this was fundamentally incorrect.

The term CNO was thereafter replaced by the term Cyberspace Operations (CO) and the terms CNE, CNA and CND were abandoned in the revised US doctrine. Three new sub-categories namely Department of Defense Information Network (DoDIN), Offensive Cyberspace Operations (OCO) and Defensive Cyberspace Operations (DCO) were introduced and superseded the earlier classification. CNE became a sub-category under OCO. The new term DoDIN signifies efforts aimed to provide information assurance on own networks and appears to share commonality with DCO at least in terms of tactics if not objectives. The old terminology of CNO, CNE, CNA and CND however still remains popular in literature and usage despite being flawed.

The confusion is amplified by use of the term "cyberspace" (cyber in short). There is no universally accepted definition of cyberspace even though it is the most commonly used slangs of our times. The 2014 revised version of the US IO Doctrine JP3-13 defines cyberspace and CO as under:

> "*Cyberspace is a global domain within the information environment consisting of the interdependent network of information technology infrastructures and resident data, including the Internet, telecommunications networks, computer systems, and embedded processors and controllers. CO is the employment of cyberspace capabilities where the primary purpose is to achieve objectives in or through cyberspace*"[5]

Above definition appears to cover the entire electronic battlespace. Whether it includes or excludes the transmission terrain covered by EW and SIGINT operations remains unclear. Also no distinction is made between military and non-military targets. A clear and concise definition helps in identification of objectives, segregation of targets and conduct of operations. It also has a direct impact on the force structure and the underlying weapons and equipment. The above definition does not appear to satisfy this criterion.

In this book we will use the term cyber warfare to signify a popular type of IW capability which targets software and hardware vulnerabilities of computational devices. By doing this we are able to maintain a clear segregation in terms of target and tactics with the other electronic terrain capabilities like EW and SIGINT. In fact it may be much better to call it "computational warfare" which future proofs the definition from any new jargon (for example the rise of artificial intelligence has already led to coining of yet another new term called "algorithmic warfare"[6] but it gets covered if we use the term computational warfare).

Sub-components and Terminology

In terms of our terrain based framework, cyber warfare (preferably called Computational Warfare) will comprise of the following operations:

- **Acquisition of Electronic Information Resident in Computational Devices.** This is popularly referred to as CNE. It involves operations to acquire information resident in computational devices by exploiting software/hardware vulnerabilities. It may be more appropriate to call it "Exploitation of Computational Devices (ECD)".
- **Attack on Electronic Information Resident in Computational Devices.** This is popularly referred to as CNA (or OCO as per latest US terminology). It comprises of operations to degrade, manipulate, disrupt, deny or destroy the functionality of computational devices or the data resident therein. The attacks again rely on exploiting vulnerabilities in software and hardware. It may be more appropriate to call it "Attack on Computational Devices (ACD)".
- **Defense of Electronic Information Resident in Computational Devices.** This is popularly referred to as CND (or DCO in US parlance). It involves operations to protect computational devices against acquisition and attack operations. Defensive operations include hardening of devices, monitoring, analysis, early detection and response to intrusions and attacks. It may be more appropriate to call it "Defense of Computational Devices (DCD)".

As already highlighted, computational devices comprise not just computers but the entire rapidly expanding universe of gadgets which use electronic hardware and software and which are vulnerable to computational attacks. Thus it includes smart phones, switches, routers, smart TVs, autonomous vehicles, robots and artificial intelligence based systems etc. All types of software like operating systems, software applications, databases and Artificial Intelligence (AI) algorithms also get included. Electronic networks are covered to the extent of hardware/software vulnerabilities of devices; however attacks which do not exploit hardware/software vulnerabilities (for example jamming, tapping of optical cables) are excluded. Though the dividing line is often thin and overlaps exist; having a precise definition helps us maintain a clear distinction between cyber warfare and the other capabilities in the electronic realm. Examples given below will help clarify some of the grey areas:

- *Insider Attacks or Computational Devices.* Insider attacks are one of the most common, dreaded and damaging type of cyber warfare threats. They come in many flavors. The first flavor is where an insider uses his/her authorized access to conduct acquisition or attack operations. Innumerable examples exist with Edward Snowden and Bradely aka Chelsea Manning being the most famous. The second flavor is where malicious software is unwittingly executed on a network by an unsuspecting insider. The Stuxnet worm which damaged the Iranian nuclear program is for example believed to have been introduced into the computational systems via infected USB flash drives. The third flavor of insider attack is phishing which involves sending malicious attachments via e-mail which are executed by an unsuspecting insider. It is one of the most widely used methods of conducting cyber warfare. What is interesting to note here is that the first example is a classic case of espionage/sabotage which is purely an intelligence operation and involves no malicious payloads. The second example is a mix of espionage/sabotage (planting infected USB flash drives in a manner that they reached insiders working in the nuclear complex) as well as cyber warfare (use of malicious payloads). The third example does not involve any espionage or sabotage but does use cyber warfare (use of malicious payload). A lay person will club all three examples above into cyber warfare. A person who understands IW and the information terrain will treat the first case as a pure intelligence operation, the second as a synergistic combination of intelligence and cyber warfare operations, and the third one as a pure cyber warfare operation which does not involve any other capability. Only if one has such clarity can force structures be evolved and operational synergy amongst the various capabilities can be achieved.
- *Attack on Radars.* Radars are computational devices. They transmit pulses (or waveforms) and then use the reflected signals to calculate the location and velocity of targets as also track them. Two common types of attacks used by aircrafts against such radars are Range Gate Pull Off and Velocity Gate Pull Off. In these attacks the jammer on the aircraft first starts transmitting duplicate pulses (or waveforms) similar to the reflected ones that the radar is receiving off the aircraft, albeit at a slightly higher power so that the radar locks onto and starts processing the duplicate signal. The parameters (like pulse width, pulse

repetition frequency) are then gradually varied to deceive the radar and pulling it off the actual target. Radars likewise have defensive techniques to counter such attacks. All these are classic EW technique involving ESM (target acquisition), ECM (jamming) and ECCM (counter-measures to overcome jamming). The underlying techniques and equipment have evolved since the days of the Second World War. When we call upon cyber warfare experts to analyze this scenario, they describe it as introduction of malicious payloads into the radar processors and a classic EW technique is suddenly hijacked and turned into a cyber attack. The key lies in discriminating the nature of the malicious payload. If it is a piece of software code then it is cyber (computational) warfare, if it is a waveform then it is EW. Many a casual expert gets fooled by over imaginative descriptions.

- *NAVWAR.* Navigation warfare which involves jamming and spoofing (deceiving) Global Navigation Satellite Systems (GNSS) like the US GPS, Russian GLONASS, Chinese BeiDou and European Union's Galileo is an increasingly vital IW capability possessed by most modern armies. The most well known examples of NAVWAR is perhaps Iran's downing of the US RQ-170 UAV in 2011. This IW operation essentially involved two steps. First was jamming of the UAV's link with its home base thus triggering its default setting where the UAV heads back to its home base. The second step was transmission of a deceptive GPS signal containing false coordinates (spoofing) which was more powerful than the actual GPS signal forcing the UAV to land at the wrong (desired) location. Once again NAVWAR is a pure EW operation, however if we search on the Internet we are more likely to find articles describing this as a feat of cyber warfare which is completely misleading. No malicious software payloads were delivered or executed. The vague and overarching definitions of cyber warfare and cyberspace propagated by many over enthusiastic but ignorant authors has led to so much confusion in the minds of decision makers that it deserves to be treated as a form of deception in itself. It has also led to the fallacy that EW and cyber warfare is converging. They are not.

The above examples make the distinction between EW and cyber (computational) warfare amply clear. If it is a piece of software code targeting the computational device then it is cyber (computational) warfare, if it is a waveform then it is EW. Replacing "cyber warfare" with "computational warfare" may resolve some of the confusion but it still requires a discerning mind.

Despite robbing cyber warfare of some of its sheen; a very complex and impressive matrix of military and non-military targets lies within the realm of cyber (computational) warfare. Computational devices today are so omnipresent that the attack surface is almost infinite and contiguous. Military and non-military targets are often juxtaposed, overlapping and indistinguishable. This in turn leads to the distinction between peace and wartime operations as also military and non-military targets becoming increasingly blurred. Cyber warfare attacks on critical infrastructure like telecommunications, banking and power grids are a glaring example of how non-military targets can be engaged to produce a decisive impact on military operations.

Above peculiarities also lead to multiple actors like hackers, non-state actors, private and government agencies as well as the military joining battle in a common arena. One possible way to understand the conundrum is to understand cyber warfare from the perspective of wartime and peace-time operations and the various actors that conduct operations in this sub-terrain.

Cyber Warfare during War and Peace

A unique facet of the information battlespace is that it is a global ecosystem comprising of pure military targets as well as targets lying in the non-military domain (like government installations, research facilities, mass media, civilian space assets, industry, telecommunications, finance, transportation, power grids etc). Many of these become legitimate targets only upon outbreak of hostilities provided they have been subjected to surveillance (acquisition) and mapping as targets during peace. This applies to all IW capabilities, but is most pronounced in case of cyber warfare due to the elephant in the room called Internet.

Today almost everyone and everything is connected over the Internet through computational devices. The technical implementation of Internet is such that it allows dissociation from real identities and confers a degree of anonymity. This relative anonymity combined with the absence of internationally agreed laws and the shared, omnipresent, borderless, distributed and instantaneous nature of the Internet result in cyber acquisition and attack operations becoming a weapon of choice for employment during peace as well as war and by actors cutting across organizations and hierarchies. There are innumerable different actors (information warriors) inhabiting this terrain each conducting their own legitimate or illegitimate operations to fulfill a vastly diverse set of missions. The underlying tools and techniques are identical in all cases but the terminology differs.

Cyberspace operations conducted during peace therefore do not get categorized as cyber exploitation, attack or defense. Peacetime cyber exploitation for example is deemed a sub-set of intelligence operations (SIGINT) because exploitation of another nation state would otherwise be deemed as unlawful. In yet another linguistic twist, cyber exploitation when carried out with state sanction against own citizens maybe called "legal interception" or "surveillance".

Similarly any peacetime cyber-attacks against foreign targets would invariably be construed as violation of national sovereignty and may even be construed as an act of war. It is therefore better to call it cyber retaliation (ie an offensive operation in response to provocation) though the term is not very popular.

Similarly cyber defense during peacetime is often called cyber security. Defensive operations undertaken by non-state actors and criminals are however construed as a crime. Similarly cyber exploitation or attack when carried out by non-state actors gets called cyber-crime or cyber terrorism.

A diagrammatic depiction of the war and peacetime sub-divisions of cyber warfare is shown below. The tools and techniques used under each sub category are identical:

Figure 3.3: Sub Components of Cyber Warfare

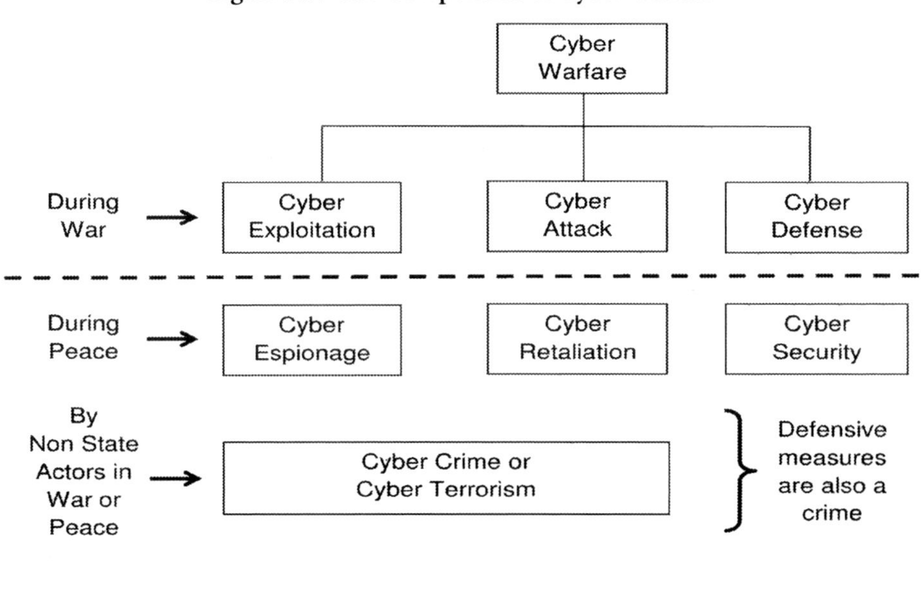

With computational devices becoming part of every system and cognitive perceptions being shaped by Internet based social media; cyber warfare shares major synergy and overlap with all the other core IW capabilities.

Doctrinal Dilemmas

The evolving and overlapping nature of IW capabilities poses several dilemmas when it comes to formulating operational doctrines and creating force structures. Four basic types of dilemmas can be discerned. These are outlined below:

- **The Tactical Diversity Dilemma**. While information itself is a continuous battlespace; there are many types of sub-terrains within each of the three primary types (cognitive, physical, and electronic). New ecosystems are constantly emerging particularly in the electronic realm. The tactics associated with each type of terrain/sub-terrain and the requirement in terms of planning/conduct of operations; equipment and training are very different. This expanding diversity of terrain and tactics results in a constant dilemma as to what operations should be combined into capabilities or which ones should be left out (and to what extent). Also many IW capabilities are strange bedfellows though they all are now converging towards the electronic terrain.
- **The Historical Dilemma**. History and evolution poses yet another dilemma. While information has acquired battlespace status only recently after having crossed an inflection point; the art and science of information operations has evolved for centuries starting from the cognitive realm, later joined by the physical and more recently the electronic. Thus organization, procedures, drills and doctrines have steadily evolved over decades and many capabilities (like intelligence and SIGINT) are so well entrenched and indispensible part of military hierarchy that revamping them is not easy.
- **The Ubiquity Dilemma**. Another big dilemma is because of the ubiquitous nature of information. Today every commander, every soldier and every system generates (and requires) information. The information battlespace is therefore constantly expanding and now stretches far beyond traditional military areas of responsibility (viz Internet and social media). This makes it difficult to consolidate the IW capabilities, assign responsibilities or impose restrictions. The

dilemma is similar to what is posed by the constantly increasing number of air space users. The second problem posed by the ubiquitous nature of information and the vastness of the battlespace is that multidisciplinary organizations can become unwieldy, unmanageable and wasteful if we attempt to combine too many capabilities across different information terrains. Creating mixed inter-departmental organizations with both military and non-military components likewise poses major administrative and functional challenges.

- **The Command and Control Dilemma**. Information being power, no commander or head of organization wants to lose control of integral IW resources and run the risk of being deprived of timely information or intimate support. Yet, information being a distinct and continuous battlespace requires that command and control be unified, and a single hierarchy is made responsible for all operations, planning, equipment, organizations and training. Catastrophic failures like the inability to detect the Kargil intrusions of 1999 in India and the terrorist attack of 2001 in the US highlight the need to unify command and control, particularly of the information acquisition operations. Yet it is easier said than done.

We can clearly see that all the four dilemmas cannot be resolved simultaneously. It is extremely difficult to combine all types of information operations into a single monolith or create one single hierarchy and entrust it with the entire IW capability.

For example, some of the capabilities may need to be left out of the purview of IW due to the tactical diversity dilemma. Typically IW doctrines leave out capabilities pertaining to physical information (like HUMINT), the inter-battlespace capabilities (like physical destruction, espionage and sabotage) or the ones which are multi-disciplinary or inter-departmental (like intelligence). The resultant simplification comes at the cost of fragmentation of the information battlespace and a loss of warfighting capability. The same is true if we try to combine them regardless of synergy.

Historical and ubiquity dilemmas require multiple organizations to handle niche IW operations (particularly acquisition) to support their respective mandate. It is for this reason that every country has an external and internal intelligence agency independent of the military. Other ministries like foreign affairs, revenue and finance may also have their own independent information acquisition organizations. Within armed forces the Army, Navy, Air Force, Marines and Coast Guard have their independent IW setups. The mission and operational requirements of these organizations are extremely diverse and merging all the IW capabilities can make the organization unwieldy, inefficient and unresponsive to unique requirements. The penalty of allowing multiple organizations comes in the form of lack of cohesion, duplication of effort and wastefulness. It may also result in splitting acquisition operations from attack and defense. The command and control dilemma remains unresolved.

Several approaches have been tried to resolve these conundrums, yet most modern armed forces have struggled to create IW force structures and hierarchies. Early doctrines (1990s vintage) tried to combine all information related capabilities under a single umbrella. Command and control warfare (C2W), electronic warfare, intelligence-based warfare, deception, media warfare and hacker warfare were all deemed to be part of IW. These doctrines soon ran into the dilemmas we described above and the capabilities therefore could not be fully integrated and operationalized. Some of the names also became obsolete fairly quickly.

Another popular approach has been to split IW operations into technical (essentially electronic) and non-technical (essentially cognitive) operations. This approach classifies information operations into Information Technical Operations (ITO) and Information Psychological/Influence Operations (IPO/IIO). ITO comprises of EW plus Cyber Warfare and IPO comprises of Psy Warfare and Military Deception. This approach does little to resolve the dilemmas cited above and completely leaves out the physical realm of information.

Another approach popularized by US IW doctrines was to split the IW capabilities into core capabilities, related capabilities and supporting capabilities. This complicated approach has been abandoned in the revised US doctrine but still finds mention in a large amount of literature.

Regardless of these dilemmas, the US has eventually created a Cyber Command and China its Strategic Support Force. Since the proof of the pudding lies in eating it, these organizations tell us many things about what is possible and what is not.

Before we start analyzing these force structures, it would be appropriate for us to examine the underlying weapon systems in order to gain a complete in-depth understanding of the subject.

ENDNOTES

1. US Joint Publication 3-13: Information Operations, (February 2006), Chapter-I, p.7.
2. US Joint Publication 3-12: Cyberspace Operations, (June 2018), Chapter-II, p.3.
3. US Joint Publication 3-13: Information Operations, (November 2012, Amended November 2014), Chapter-II, pp. 5-13.
4. Ibid, Chapter-I, p.4
5. US Joint Publication 3-13, op,cit, Chapter-II, p.9.
6. Sydney J. Freedberg Jr, "*Algorithmic Warfare: DSD Work Unleashes AI On Intel Data*" (Published on Breaking Defense Online Magazine Website, 28 April 2017), Web URL: https://breakingdefense.com/2017/04/dsd-work-unleashes-ai-on-intel-data-algorithmic-warfare/ accessed on 01 Mar 2018.

CHAPTER 4

Weapon Systems

Weapons are integral to any battlespace. Traditionally speaking, the main function of a weapon is to inflict damage upon a target. Thousands of different kinds of weapons exist and they can be classified based on their characteristics (viz rockets, missiles, machine guns, assault rifles, nuclear weapons, biological and chemical weapons etc), or the type of target they engage (viz anti-tank weapons, anti-aircraft weapons, torpedoes etc). Another term we often encounter is "weapon system". While bombs, rockets, machine guns and various types of missiles are each a weapon – a fighter aircraft integrates all of them into a single platform. Such platforms can be deemed as a weapon system. An aircraft carrier can be visualized as a still higher level of integration and maybe deemed as a system of systems.

Some weapons like machine guns and grenades are used both for attack and defense; however in most cases it is possible to characterize weapons or systems based on whether they are offensive or defensive. Aircrafts, attack helicopters, missiles, tanks and artillery for example are used predominantly for attack whereas mines, anti-aircraft, anti-tank and missile defense systems are predominantly used for defense. Almost all fighting formations are equipped with a mix of offensive and defensive weapons depending upon their role; however major offensive and defensive weapon systems are usually held and operated by different formations and units. Armor protection and field fortifications are not deemed to be weapons despite their being vital to defense. A bullet proof vest is similarly not deemed as a defensive weapon despite it being as vital to survival as an assault rifle.

The problem of defining weapons and weapon systems multiplies dramatically when it comes to the information battlespace. We can physically destroy a computer or a microwave transmitter using kinetic weapons but it is not possible to hit software vulnerabilities with a bomb or damage an electromagnetic wave by firing a machine gun at it. Cognitive perceptions can likewise not be engaged by munitions. The traditional meaning of a weapon mostly fails us in the information battlespace. The reason for this is that information is an abstract battlespace.

Targets in the information battlespace cannot be acquired (or pin-pointed) using latitudes, longitude, distances, grid-references, velocity, elevation, altitude or depth. Nor can they be damaged by kinetic energy, heat and fire. Targets in the information battlespace are instead pinpointed through a different set of identifiers like frequency, IP address, port number etc in the electronic terrain or identifiers like nationality, tribe, ethnicity, language, financial status, age and religion in the cognitive realm. They can only be damaged through compatible malicious information payloads.

The only way to define a weapon in the information battlespace is to go back to first principles and simply treat a weapon as a device, object, application or tool which can facilitate attack or defense. If it is possible to aggregate such individual weapons onto a single platform, then we may also call them as weapon systems. As we shall see subsequently, it is possible to classify weapons in the information battlespace into offensive and defensive similar to the physical battlespaces. Also major offensive and defensive information weapon systems are held and operated by different formations and units just like in the physical battlespaces.

If the above definition of an information weapon sounds too farfetched, it needs to be highlighted that the US Air Force designated six cyber tools as weapon systems in the year 2013[1]. It would come as an even greater surprise to know that these weapon systems were defensive cyber-security software suites meant to be used for vulnerability assessment, intrusion detection and cyber-attack mitigation[2]. These weapon systems are wielded by special units responsible for defending the Air Force portion of the global information grid.

With this preamble, let us look at what weapons are used by the four major capabilities we have indentified in the previous chapter.

Psy Warfare Weapons

Control of hearts and minds is the pinnacle of information warfare and it can be achieved through psy warfare. Psy warfare is conducted in the cognitive terrain and targets are the minds and thought process of individuals. Unlike the physical or electronic domain, cognitive targets can only be perceived but not pin-pointed. They can only be attacked (or defended) through cognitive weapons and there exist a large number of objects, systems and tools designed to bombard the human mind with information payloads configured in a manner to disrupt, degrade or manipulate the cognitive process. In fact one of the most discernable characteristic of the information age is that we are all subjected to a never ending stream of information, all designed to influence the way we think and act. It is also a fact that since we are consuming increasing amounts of electronic information, therefore many of the weapons also make use of the electronic terrain to infiltrate into the cognitive realm. Weapons similarly exist in the physical terrain and they assume importance in active battlefield scenarios where electronic vectors like Internet and TV are often not available. A brief description of the weapon systems used for conduct of psy warfare is as under:

- **Weapons in Cognitive Terrain**
  - Direct communication viz political dialogues, diplomacy
  - Public addresses
  - Protests and demonstrations
- **Weapons in Physical Terrain**
  - Posters/pamphlets/leaflets/banners/pictures
  - Newspapers and magazines
  - Books
  - Objects of daily use carrying informational payloads
- **Weapons in the Electronic Terrain**
  - Radio
  - Television
  - Internet based platforms
    - Social media and social media engineering tools
    - Blogs, websites, forums, podcasts
  - Data analytics software
  - Artificial intelligence and bots

Intelligence Weapons

As already explained; Intelligence is a vast discipline comprising of HUMINT, SIGINT, GEOINT, MASINT, IMINT and OSINT etc as well as espionage and sabotage. Each of these disciplines employs different tools and techniques to acquire or attack (degrade, manipulate, disrupt, deny and destroy).

Countermeasures against these form part of defensive weapons. A brief description of the weapons used is as under:

- **HUMINT**. The weapons in this case are human agents.
- **SIGINT**. Being an amalgamation of all electronic acquisition operations; SIGINT uses a vast array of application specific electronic receivers and sensors to tap into radiating transmission, fiber optic cables, telephone cables and telecommunication networks including Internet. Special software suites for packet inspection, data analytics and aggregation/collation of information are likewise weapons used as part of this capability. The use of supercomputers for cryptologic operations is likewise common.
- **IMINT**. Satellites, aerial reconnaissance aircrafts and UAVs are some of the weapon systems used by IMINT. GIS software suites are essential to IMINT.
- **MASINT**. Electromagnetic, optical, seismic, acoustic, biological, radiation, and chemical sensors are the commonly used weapon systems and these are deployed on land, sea and air.
- **OSINT and FININT**. Weapons used are similar to SIGINT and HUMINT.
- **Espionage and Sabotage**. Agents and spies are the weapon systems used for espionage and sabotage. They in turn may use a variety of special gadgets and devices to facilitate conduct of operations which may also be deemed as weapons.

EW Weapons and Systems

There are a large number of EW weapons and they can be grouped in many ways based on common attributes. It is also possible to organize them into larger systems. A simplified overview of the weapons, systems and the various categories into which they can be classified is as under:

- **EW Weapons Based on Type of Operation.** Three obvious categories exist corresponding to the acquisition, attack and defense operations. Brief details are as under:
  - **ESM Systems (Acquisition Operations)**. ESM systems comprise of receivers which scan the electromagnetic spectrum and pick up active frequencies on which transmissions are taking place and

discern their geographical location. Hostile frequencies are identified and earmarked for monitoring or jamming.

- **ECM Systems (Attack Operations)**. Active ECM systems target the receiver by transmitting high power signals. The receiver is unable to process the original transmission because the jamming signal is more powerful. Many different jamming techniques exist and many types of target specific waveforms can be used. Passive ECM systems like chaff and decoys distort the signal reaching the receiver. DEWs are another distinct class of weapons.
- **ECCM Systems (Defense Operations)**. Most modern radars and communication equipment are designed to use spread spectrum techniques like frequency hopping. Use of ECCM hardware does not completely eliminate the possibility of acquisition or attack but it increases the chances of survival (ie reduces the probability of acquisition as well as requires higher jamming power and special type of jamming techniques or waveforms to attack). It is also possible to design receivers in such a way that they are able to detect ECM attacks and take evasive action (viz switch frequencies or filter out/suppress the spurious signal).

- **EW Weapons Based on Type of Target**. There are application specific EW systems designed for specific types of targets. The reason for this is that propagation characteristics, antenna systems and the electronic circuitry (detectors, demodulators, power amplifiers etc) are different for different frequencies. Even if the frequency is same, different types of systems maybe required because specialized waveforms are required to be transmitted in order to jam a receiver. A simple analogy is that a 2G mobile handset cannot demodulate a 4G signal because even though the frequency bands overlap, the underlying hardware is different. Some of the commonly encountered target specific EW weapon systems are as under:
  - **Counter RCIED EW Systems (CREW)**. These systems transmit jamming waveforms which either cause premature detonation of the IED or else block the radio signal used to detonate it. These systems are typically vehicle mounted or manpack.
  - **HF, VHF and UHF EW Systems**. These systems target HF (both sky and ground wave), VHF and UHF links. Many sub-types of

EW systems exist within this broad category because of the wide range of applications. For example the UHF band is used by microwave links, television broadcast, GPS, cell phones, hand held satellite phones, Wi-Fi, bluetooth and even UAVs. Each one of these applications requires a different type of EW equipment and antenna configuration. HF, VHF, microwave, GPS and cellphone jammers are thus all unique and different from each other. It is also possible to have multiple configurations like manpack, ground based, vehicle mounted as well as airborne systems depending on the range, power and mobility requirement.

- **Anti-Satellite (ASAT) EW Systems.** These EW systems are specifically designed to target space based systems. They may target specific satellite based applications (like point to point links or VSAT networks) or the satellites themselves (viz their onboard sensors, command, control, tracking and telemetry). Once again many sub-types of equipment will exist for different bands of operation (viz C, X, Ku, Ka etc) and these maybe static or mobile.
- **Radar EW Systems.** These are a distinct class of EW equipment designed to jam or deceive radars. Since radars operate across a wide range of frequencies, the EW systems to target these are also different for different types of radars.
- **Ground to Air EW Systems.** This is another distinct class of EW systems specifically designed to target electronic systems and sensors aboard UAVs, UCAVs and airborne platforms like attack helicopters and AWACS.
- **NAVWAR Systems.** These EW systems are designed to jam and spoof GNSS signals. They are again application specific and a system will be able to target GPS, GLONASS, BeiDou etc only if it has been designed to transmit the applicable waveform.
- **Platform Centric EW Suites.** The main purpose of these suites is to protect high value platforms like aircrafts and ships. These EW suites are very closely integrated with other sub-systems and sensors on the platforms both in terms of hardware and software. The tasks and responses are often automated. EW suites essentially augment the stealth and self-protection of their platforms wherein the aircraft or ship can evade detection through jamming and/or

deceiving enemy radars as also avoid being hit by incoming projectiles through jamming and/or deceiving their guidance and homing mechanisms. Systems use both active (jamming/ deception) and passive (chaff/decoys) techniques. Active and passive EW systems are thus yet another distinct sub-category.

- **EW Systems Based on Platforms.** Another method of classifying EW systems is based on the platforms on which the equipment is carried. This essentially leads to the following classification:

  - **Manpack EW Systems.** These are carried by individual soldiers. They have low weight and can be carried cross country but suffer from limited range and jamming power due to limitations of antenna size and electrical power. Small antennas have poor acquisition sensitivity and high power jammers (transmitting say 500 Watts or 1 KiloWatt power) require big generator sets to power them and cannot be carried by individuals. Handheld EW systems like anti-drone guns or disposable GPS jammers are an application specific sub-category of manpack systems. Manpack EW systems cannot cover all the applications or the entire spectrum.
  - **Ground Based EW Systems.** These systems can be static or mounted on vehicles. They are not severely constrained in terms of size, weight, volume and electrical power though combining too many applications and frequency bands on vehicle mounted entities can pose a problem in terms of the number of receivers, analytical posts and antennas that can be accommodated. Increasing the vehicle size results in a penalty in terms of camouflage and maneuverability, especially in mountains. Cross country mobility is often a mandatory requirement and systems must be mounted on appropriate platforms. Space is however a constraint in case of systems mounted on armored fighting vehicles. Availability of line of sight to the target is a vital consideration regardless of platform. The biggest advantage of ground based EW systems is that they can close in with the enemy and operate on a 24/7 basis provided logistics (fuel, food, water) and physical protection are catered to. There are distinct tactics associated with employment, deployment and integration of these systems.

- **Airborne EW Systems.** Airborne EW systems are somewhat similar to ground based EW systems albeit they are mounted on aircrafts, helicopters, UAVs and also aerostats. It must be noted that this is different from EW suites which are already integrated into the platform for stealth and self-protection. Such airborne EW systems are either specially modified aircraft designed for EW operations or else carry an additional EW payload. The biggest advantage of such airborne systems is the extended ranges (greater line of sight) they enjoy due to their elevation wherein they can acquire (or jam) targets from a much greater distance and jam them using lesser power. Line of sight constraints mostly do not apply and they can be easily and quickly maneuvered to the area of operations. These systems are however not a substitute for ground based systems because the operations are limited by the flying time and can rarely be carried out 24/7. Airborne EW systems are also constrained in terms of space, weight and onboard power which can be made available. Considerable miniaturization and aerodynamic integration effort is required because unlike ground based systems, they cannot hoist antennas on masts. The overall cost of such systems is high.
- **Sea Based EW Systems.** Ships are not as severely constrained in terms of space or power as airborne platforms. However ships operate in high seas and their main concern is stealth and self-protection. Accordingly all modern ships have integrated EW suites (including jammers) but do not carry out continuous jamming operations like land or airborne EW systems. They do not transmit electromagnetic energy unless absolutely necessary. Amphibious operations are one area where ship based EW systems may be deployed to engage ground based targets. Passive surveillance (ESM), particularly of satellite transmissions is yet another area of application. Satellites (particularly military) have a well defined geographical footprint which maybe inaccessible to land based systems but ships may travel to reach within it. Use of submarines to acquire information from undersea optical fibre cables (or destroy them) is another niche application but it is strictly speaking an intelligence capability (SIGINT/sabotage) and not EW since the transmission is non-radiating.

- **Integrated EW Systems.** Each of the various types of EW weapons (viz manpack, ground based, airborne or ship based) described above can be employed to support operations in a standalone manner. This however has a major limitation. It is akin to each weapon being allowed to choose targets (frequencies, links and applications) which it may engage without any mutual coordination or control. This leads to duplication (viz the same link being monitored or jammed by multiple systems) and the effort must be synchronized (integrated) across the entire range of targets and amongst the available weapon systems in order to obtain maximum and timely impact. One of the fundamental principles of EW operations is that its strength lies in integration. Three things are required – firstly intra-communication systems to network all entities, secondly a unified database, and thirdly an integrated command and control entity which can dynamically assign targets and control the EW operations by leveraging the database. Integrated EW systems (IEWS) incorporate all these three attributes wherein individual EW platforms (entities) are linked over a network to a higher control station located in depth. The ESM data captured by individual entities is fed into a centralized database held at the control station which then analyses, prioritizes and assigns ESM and ECM tasks to each forward entity. IEWS thus comprise a grouping of many diverse entities which can cover different frequency bands and multiple targets. Such a collection of centrally controlled entities is often called as an IEWS block. IEWS blocks can be organized into companies and battalions to support tactical formations. Control stations are often co-located with formation headquarters and support offensive operations, situational awareness and integration with kinetic means. IEWS operations and tactics are a complex and specialized discipline.

Cyber (Computational) Warfare Weapons and Systems

The computational terrain has its own unique set of offensive and defensive weapons. There are no classifications which can be derived based on how they are carried (viz manpack, vehicular, airborne etc). There are also no limitations of range, line of sight, weight, volume or power. Classifying cyber weapons thus requires different parameters.

One example is "zero day vulnerabilities" and their weaponisation like the eternal blue exploit developed by the US National Security Agency (NSA). It would be appropriate to deem each such malware which exploits a particular zero day vulnerability (of an operating system or application) as a weapon. However there is no limit to the number of such vulnerabilities (and weapons) because the software ecosystem keeps expanding. Also the life of such weapons is limited till such time that the vulnerabilities are detected and patched. This makes it difficult to classify any of these as a distinct class of cyber weapons.

We can also classify weapons based on parameters like operating systems (viz Windows, Linux, Android, iOS). It is also possible to group cyber weapons based on tactics and we can classify them as viruses, worms, trojans, spyware, keyloggers, rootkits, bots, ransomware etc. There are also customized weapons which are specially crafted for use against a specific target. The stuxnet worm used to damage the Iranian nuclear reactors is one of the best known examples. It is thus also possible to group weapons based on the targeted hardware.

The above methods of classifying cyber weapons is however inefficient because there is a requirement of integrating the capability and synchronize the effort across various types of targets and operations (i.e. systems). This leads us into the realm of integrated cyber warfare suites (or frameworks) which can deliver multiple types of malicious payloads onto the target and ensure continuity of operations. Defensive suites can likewise be used to protect against threats and can be constantly upgraded with new definitions and patches to protect against emerging threats.

Customized suites which allow us to integrate the offensive and/or defensive operations and carry multiple individual weapons (exploits) cutting across tactics and hardware are therefore a more appropriate candidate for being termed as a weapon system rather than a mere tactics, operating system or hardware based grouping. Many such specialized offensive and defensive suites exist and most cyber warfare units have their own customized inventory of these. It is not possible to list all of these weapon system however some representative examples are given below to drive home the concept of cyber weapon systems:

- **Metasploit Framework.** Metasploit framework is an open source vulnerability assessment and penetration testing software suite. The intended purpose of all penetration testing software is to attack the system with a view to identifying vulnerabilities. Obviously it can also be used to carry out acquisition and attack operations. Metasploit framework has tools to scan targets in a network and identify vulnerabilities. It also has hundreds of different exploits covering multiple operating systems which can thereafter be selected and delivered via configurable payloads. Anti-forensic and evasion tools are also inbuilt into the framework thus allowing the attacker to remain anonymous[3]. The product is constantly upgraded with new exploits and attacks.

- **Kali Linux**. Kali Linux is a customized linux distribution designed for digital forensics and penetration testing. It has a large collection of offensive tools and software programs[4]. Once again multiple types of operations can be launched. The intent of operations maybe to test the robustness of defenses and initiate corrective measures, exfiltration of information or to cause damage to the target. The product is again constantly upgraded with new exploits and attacks.
- **EnCase**. EnCase comprises a suite of defensive software products for detection and mitigation of threats as well as carrying out forensic analysis to recover evidence of attacks[5]. It can be deemed as a defensive weapon system.
- **Cyberspace Vulnerability Assessment/Hunter (CVA/H)**. CVA/H is a defensive suite used by the US Air Force Cyber Command and deemed to be a weapon system. The description of its features states that "*the CVA/H weapon system is designed to identify vulnerabilities and provide commanders with a comprehensive assessment of the risk of existing vulnerabilities on critical mission networks. It is functionally divided into a mobile platform used by operators to conduct missions on-site or remotely; a deployable sensor platform to gather and analyze data; and a garrison platform which provides the connectivity needed for remote operations as well as advanced analysis, testing, training, and archiving capabilities. Additionally, the Hunter mission focuses on the capability to find, fix, track, target, engage and assess the advanced persistent threat.*"[6]

It is obvious from the above examples that while individual pieces of malware targeting different operating systems or hardware are indeed weapons, new ones are constantly appearing and this is an inefficient way to classify cyber weapon systems. Instead the specialized offensive and defensive suites/ frameworks are much more appropriate to be deemed as cyber warfare weapon systems. The EternalBlue exploit developed by the US NSA and which was leaked and thereafter used in the worldwide WannaCry ransomware attack has for example been fully integrated into the Metasploit framework[7].

Conclusion

What we gather from the above discussion about weapon systems is that each of the capabilities has a unique set of weapon systems to inflict damage upon enemy information (or to defend own information) in their respective terrain/ sub-terrain. It remains a matter of choice whether we wish to call them weapon systems (as the US armed forces do) or continue to call them as equipment, tools, techniques, software etc.

It is also evident that IW weapon systems are extremely diverse in terms of their design, tactics and employment concepts. This reinforces the belief that the IW capabilities are often strange bedfellows and while these weapons may be used in mutual support and synchronization, they remain unique and can be clearly differentiated.

In the next chapter we will try to examine how IW capabilities (and underlying weapons) integrate with kinetic means and also amongst each other. We will also cover the conduct of IW operations at various levels of war and miscellaneous aspects like training and capability building.

ENDNOTES

1. Defencesystems.com website, "*Air Force designates six tools as cyber weapons*" (Published on 11 April 2013), Web URL: https://defensesystems.com/articles/2013/04/11/air-force-cyber-weapons.aspx accessed on 01 December 2017
2. RC Porter, "*Air Force Details 6 Cyber Capabilities That Are Now Weapon Systems*" (Published on Fortuna's Corner website, 18 Aoril 2013), Web URL: https://fortunascorner.com/2013/04/18/air-force-details-6-cyber-capabilities-that-are-now-weapon-systems/ accessed on 01 March 2018
3. Rapid7 website, "*Metasploit Pen Testing Tool*", Web URL: https://www.rapid7.com/products/metasploit/download/editions/# accessed on 30 June 2018
4. Kali Tools website, "*Kali Linux Tools Listing*", Web URL: https://tools.kali.org/tools-listing accessed on 30 June 2018
5. Guidancesoftware website, "*EnCase Endpoint Security*", Web URL: https://www.guidancesoftware.com/ accessed on 30 June 2018

6. Website of the US Air Forces Cyber, "*Cyberspace Vulnerability Assessment/Hunter*" (Published on 22 March 2017), Web URL: http://www.afcyber.af.mil/About-Us/Fact-Sheets/Display/Article/1186672/cyberspace-vulnerability-assessment-hunter-weapon-system/ accessed on 01 March 2018
7. Ionut Arghire, "*NSA's EternalBlue Exploit Fully Ported to Metasploit*" (Published on the Security Week website, 16 May 2017), Web URL: https://www.securityweek.com/nsas-eternalblue-exploit-fully-ported-metasploit accessed on 01 December 2017

CHAPTER 5

Integration with Kinetic Means and Conduct

Military Objectives of Information Warfare

Broadly speaking the objective of IW is to achieve information superiority and denying or manipulating the information available to adversary. In simple words this implies reducing the fog of war for own forces while enhancing it for the enemy. This however is over simplistic and does not fully capture the effects that can be obtained by applying IW in the cognitive, physical and electronic realms. Some of these effects which can be orchestrated in the information battlespace are summarized below and many more exist:

- Impact enemy morale and intimidation
- Deceive enemy about own intent and capability
- Gain covert access to enemy information (plans, capabilities, disposition and intent)
- Impose tactical delay
- Influence non-combatant and neutral organizations to support own military objectives and incite favorable response from friendly/neutral nations
- Gain situational awareness and early warning
- Monitor progress of own and enemy operations
- Carry out post strike damage assessment
- Deny use of critical infrastructure
- Disrupt command, control, communications, computing and networking
- Corrupt data/information leading to operational failures
- Deny/spoof navigation, position and timing signals
- Degrade enemy surveillance, early warning and target acquisition
- Suppress or render ineffective weapons and systems relying on communication, networking, computing or navigation
- Denial of space based assets

- Physical destruction (viz EW attacks on UAVs, computational attacks on embedded control systems, sabotage, or DEWs)
- Stealth and self protection for weapons platforms and systems
- Impose restrictions on employment of certain classes of weapons particularly nuclear, biological and chemical
- Suppress and degrade insurgent and terrorist organizations

A large number of targets exist within each sub category and a close synergy between operations in the cognitive, physical and electronic terrain yields the best results. The effects are further amplified if synergized with operations in the other battlespaces of land, sea, air and space. IW operations that are confined to a single terrain or not synergized with other battlespaces will yield sub-optimal results.

IW at Strategic, Operational and Tactical Level

The purpose of war/conflict and use of military force is always "political" in nature. The military objectives are therefore selected in a manner as to further the political aims. IW operations likewise flow from the military objectives and do not exist in isolation.

It is also necessary to understand that perspectives of war vary at the strategic, operational and tactical levels. From the strategic perspective, war is a continuum to secure national policy objectives and national security goals[1]. The end state of any war or conflict must therefore provide the desired political advantage. Also, at the highest level, the Armed Forces are just one of the elements of national power in addition to leadership, governance, internal security, social cohesion, scientific and human capital, natural resources, economy, foreign affairs and diplomacy[2]. Information is the common life blood that feeds all these vital constituents and it is therefore logical that each of these elements of national power also contributes to offensive and defensive IW in its own unique manner. IW operations at the strategic level must therefore first and foremost foster synergy of effort with the other non-military elements of national power. Feeding accurate and timely information about ongoing operations, assimilating information received from other elements/battlespaces and formulating these inputs into actionable plans and directives is therefore the main IW task at the strategic level. Information assurance is critical to the success of such actions. In addition to this, IW attacks against critical infrastructure, strategic deception and perception management are some of the key operations that get directly orchestrated and conducted at strategic level and these must be synchronized with theatre level plans. Defending against similar enemy attacks is another vital task at the strategic level.

The operational level translates abstract strategic aims into achievable tactical objectives through the design, organization and conduct of major operations. This requires coordinating the effort of land, air and naval forces in time and space. Once operations commence, the operational level must ensure uninterrupted aggregation of information about the multiple tactical engagements taking place throughout the theatre. Continuous availability of the command and control network is a vital task. Use of own IW resources to deny similar aggregation capability to the adversary comes next in the list of IW objectives at the operational level. Synchronization of IW operations and resources amongst land, air and naval forces is the third important task at the theatre level. Allocation of theatre level IW assets (viz airborne jammers) to support key tactical operations is another vital task. Applied correctly and at decisive moments, theatre level IW assets can alter the course of tactical engagements.

The main focus at the tactical level is optimizing the application of military forces to achieve specific objectives. This level focuses on the application of combat power to defeat the enemy at a particular time and place using firepower, maneuver and combined arms synergy. Since electronic surveillance, networking, shared situational awareness and self-synchronization have assumed increasing importance in the modern digital battlefield; the use of EW for jamming of enemy radars, communications and sensors is the most important IW activity that takes place at the tactical level and can act as a decisive force multiplier. IW operations, particularly EW, can help identify and track time sensitive targets and fleeting opportunities at the tactical level and facilitate their engagement with kinetic weapons. They can also help in damage assessment post kinetic strikes. Similarly, kinetic strikes can degrade enemy's communication nodes and terrestrial networks thus forcing the enemy to resort to wireless communications which are vulnerable to EW attacks. Integration of IW with kinetic means is therefore another vital function at the tactical level. Employment of IW operations to deceive or gain local tactical advantage may also be possible. Lastly defensive operations to protect own Command, Control, Communications, Computers, Intelligence, Surveillance and Reconnaissance (C4ISR) capability are equally important.

Integration of IW with Kinetic Means

When we talk of kinetic means it includes weapons in all the other battlespaces including nuclear weapons. Likewise IW includes all the core capabilities cutting across the entire information terrain. Integration refers to maximizing their individual potential through synergistic application at the strategic, operational and tactical level. Two basic forms of integration can be envisaged:

- **Replication of Effects.** In this type of integration IW and kinetic means are used as an alternative to each other. Such options allow commanders to conserve resources or else to engage the adversary on multiple fronts.
- **Amplification of Effects.** In this case both IW and kinetic means are employed in tandem to increase the effectiveness of attack or defense operations. These can be further subdivided into:
  - *Sequential Application.* In this case IW capabilities and kinetic means are applied in succession one after the other. For example IW operations (particularly EW and intelligence) can discern enemy plans and disposition allowing kinetic weapons to thereafter engage targets with greater precision and lethality thus increasing their effectiveness. It is similarly possible that kinetic means are applied first causing attrition in the information battlespace and thereby increasing the effectiveness of EW, Cyber and Psy warfare.
  - *Concurrent Application.* In this case IW operations and kinetic means are applied simultaneously. One such example is where EW can discern targets through electronic acquisition and guide kinetic platforms (viz fighter aircrafts) onto targets in real time. Another scenario is where kinetic and IW attacks are launched in tandem leading to greater confusion, delay and attrition. It is also possible to use IW capabilities to suppress enemy surveillance and command networks so that kinetic platforms can execute their mission without coming under effective fire and return back safely. Combining psy warfare with kinetic means can likewise pay rich dividends.

How such integration plays out at the strategic, operational and tactical level is tabulated in the target engagement matrix below. Both replication and amplification effects can be discerned:

- **Strategic Level**

Targets	*Kinetic Operations*	*IW Operations*
Critical Infrastructure (Satellites, financial hubs, power plants, Internet and tele communications)	Physical destruction of critical infrastructure reduces enemy's war fighting potential	Cyber warfare and EW can likewise cripple critical infrastructure without loss of civilian life
Nuclear Weapons	Credible second strike capability and missile defense systems can deter a nuclear attack	Psy warfare at the strategic level can likewise prevent or limit a nuclear conflict
Missile Defense	Kinetic weapons can intercept and destroy incoming long range missiles	Cyber warfare and EW can also be used to interfere with the guidance and homing mechanisms of incoming long range missiles
Leadership	Kinetic weapons can be used to take out key decision makers	IW can also be used to discredit or undermine key decision makers

- **Operational Level**

Targets	*Kinetic Operations*	*IW Operations*
Major operational installations (Airfields, ports, headquarters, logistic bases and ammunition dumps)	Air, naval weapons and missiles can be used to target such installations	EW can identify such targets and also help suppress enemy air/missile defense and early warning systems to facilitate kinetic strikes. IW resources can also facilitate post strike damage assessment
Concentration and staging areas in depth	Aircrafts, missiles and long range artillery can be used to engage such targets	Same as above
Counter mobility targets (Bridges/tunnels, railways, helipads, choke points, dams and water works)	Same as above	Same as above
Industrial Targets (Research facilities, laboratories, dual use factories and energy hubs)	Same as above	In addition to above; such assets can also be targeted directly through cyber warfare. IW assets can also help discriminate/isolate legitimate military targets from civilian ones
Mass Media (TV and Radio Stations)	Same as above	EW can be used as a non-lethal alternative to disrupt these

- **Tactical Level**

Targets	*Kinetic Operations*	*IW Operations*
Ground holding forces (while attacking) or assaulting enemy forces (while defending)	Kinetic means are used to cause attrition and facilitate own maneuver while denying mobility to the adversary	EW acts as a force multiplier by providing early warning and disrupting the adversary's ability to communicate/coordinate/call for reinforcement and reserves
Command and Control	Kinetic means can target Tactical HQs and communication nodes	ECM can disrupt the commander's ability to communicate/coordinate
Tanks and ICVs	Anti Tank weapons can target these	EW can disrupt the ability to communicate as well as deny or spoof electronic navigation. It can also identify targets for kinetic engagement
Artillery gun positions and Air Defense weapons	Kinetic weapons can be used to engage these	EW can identify targets for kinetic engagement as well as disrupt the ability of such entities to acquire targets and/or communicate and coordinate
UAVs, UCAVs and Unmanned Robotic Weapons	Same as above	In addition to detection and jamming, EW can also interfere with navigation and bring these down
Early warning radars, sensors and surveillance networks	Same as above	EW can detect as well as disrupt the ability of such entities to acquire data and/or communicate it
Adversaries Morale and Will to Fight	Heavy and sustained kinetic engagement combined with physical isolation of target diminishes the enemy's will to fight and can lead to troops abandoning the fight or surrender	Psy warfare can also demoralize the adversary and encourage desertion. Electronic isolation and inability to communicate can greatly increase the sense of helplessness

Integration Methodology

At the strategic level, the application of IW operations is a political decision with both military and non-military agencies involved in conduct. No single entity can control the entire spectrum of IW activities covering everything from diplomacy to cyber operations at the strategic level. Integration at the strategic level would thus require a multi-agency organization headed by the political executive. Intelligence sharing is yet another critical area where such a multi-agency organization has a decisive role to play.

At the operational and tactical level, land forces conduct IW as traditional C2W whereas airborne and naval platforms are more concerned with survivability and stealth. When it comes to integration, it is often in terms of allocating airborne EW assets in support of ground, air and naval operations. Discrimination of friendly targets from hostile ones, spectrum management and preventing fratricide is the second important task. Integration can be achieved through joint staff organizations, implementation of IFF techniques and protocols. Within respective branches, integration requires to be engineered at both the hardware and software level (viz. integration of EW systems with battlefield surveillance, decision support and targeting systems).

Defensive IW operations to ensure continued availability of own information grid is vital at all three levels.

Offensive and Defensive IW Operations

IW operations involve both offensive and defensive elements. Offensive IW ops degrade enemy's information and net-centricity whereas defensive IW ops safeguard one's own information and ability to leverage networks. The table below summarizes the main type of offensive and defensive operations applicable to the major types of information terrains:

Terrain	*Offensive Operations*		*Defensive Operations*
	Acquisition	*Attack*	
Cognitive	Intelligence	Psy Ops	Counter Psy Ops, Counterintelligence Ops
Physical	Intelligence	Intelligence (Sabotage)	Physical Information Security
Transmission (radiating)	SIGINT and ESM	ECM	Network security and redundancy, ECCM
Transmission (non-radiating)	SIGINT	Intelligence (Sabotage)	Network security and redundancy
Computational Devices	Cyber Exploitation	Cyber Attack	Cyber Defense

Typically offensive IW operations are carried out by specialized units whereas defensive IW operations are conducted both by specialized units as well as normal users. This is not very different from normal military operations where designated units and formations launch offensive or defensive operations in respective battlespace but every unit including logistic echelons are supposed to protect their own installation. Consider the case of cyber warfare. Offensive operations are launched by skilled hackers and defensive operations are conducted by skilled system administrators; yet every individual user must also secure their computers/data. Likewise in case of Electronic Warfare; ESM (monitoring) and ECM (jamming) require special equipment and trained personnel whereas defensive ECCM tasks are required to be performed by specialists as well as every radio and radar operator. Same is the case with Psy warfare too where specialized units conduct the actual offensive and defensive operations but all echelons are required to take measures to guard against malicious propaganda and ruses.

In the modern network-centric digital battlefield no meaningful operations can be launched if own C4ISR infrastructure cannot be accessed. The inability of Iraqi armed forces to defend their command/control networks from being suppressed by the US coalition forces during the gulf war is a stark example. It ruled out any meaningful retaliation and led to quick annihilation of the Iraqi defense forces. Clearly when it comes to the information battlespace, successful defense is as important as attack. Doctrines which deem IW as a predominantly offensive capability and ignore the defensive aspect are deeply flawed. Network security, redundancy and cyber defense is of paramount importance and deserves equal if not greater attention. Formations and units which provide communications and networking are therefore a vital part of the IW force structure.

Lastly, just like the other domains of land, sea and air – offensive and defensive operations in the information battlespace are closely intertwined with each other, yet the formations/units responsible for attack (the so called strike formations) have very different equipment and organization as compared to the defensive (holding) ones. The hackers (attacker) and the system administrators (defender) likewise may share some similarities but belong to different units and are equipped with entirely different set of software and hardware tools. The organizations and roles are not interchangeable.

IW in Counter Insurgency (CI) Operations

Military theorists consider population to be the centre of gravity in any military conflict and this is equally true in CI operations[3,4]. In conventional war; physical destruction is the primary means to influence the will of population to continue fighting. Breakdown of the populations will to fight typically leads to military capitulation followed by political negotiations or a truce. In CI operations it is not possible to inflict physical destruction upon one's own population; however the aim once again is to break the trust between insurgents and the local population so that the will to fight is diminished. Psy warfare plays a dominant role in influencing the population and bringing an end to the insurgency.

Another major difference is the time factor. In CI operations it is difficult to estimate the campaign duration, which in turn makes it hard to set time limits on major operational phases. Further, one must coordinate the military actions with civilian agencies to a much greater extent. Prolonged deployment, the need to remain engaged with civil population and the requirement to bring about normalcy imposes considerable restrictions on military IW operations.

Lastly, in CI operations the insurgents can effectively highlight their own successes as well as the failures and missteps of government forces. A key advantage for the insurgent is that their messages need not be based on factual information. Instead, these need only resonate briefly with the local population to create skepticism about the legitimacy of the government's cause in order to achieve the desired result. Government machinery including the military on the other hand is often seen as the oppressor and is on the back foot in terms of credibility. A dedicated and professionally designed Psy warfare campaign has great significance in asserting the moral superiority and credibility of state forces. Social media is assuming an increasingly important role in conduct of Psy warfare during CI operations.

Cyber warfare and EW have also gained increasing prominence due to insurgents making use of Internet and electronic communications for command and control, financing, propaganda and recruitment. Each of these activities can be targeted; however unlike conventional operations, the targets lie hidden amongst thousands of normal users and isolating them is a resource intensive task. Use of techniques like steganography and encrypted messengers pose a further challenge wherein a different type of equipment and capability is required from what is typically authorized to the armed forces for conventional operations. Another challenge comes from rapid change in ICT technology and the speed with which insurgents and terrorists are able to switch options. The state forces may not have the technical resources to target each and every new technology and platform in real time. However, one big advantage available to state forces is that they control availability of the communication and Internet infrastructure.

IW and Special Operations

IW can facilitate special operations at the strategic, operational and tactical levels. At the strategic level, IW support to special operations can include such actions as reinforcement of mission by the political leadership and containing any adverse media fallout. At the operational and tactical level EW can facilitate electronic isolation of targets and suppression of early warning elements to enable safe insertion and quick disengagement of teams. EW can also disrupt calls for reinforcements or fire support, assist in monitoring progress of operations and facilitate post strike damage assessment. Preemptive Psy warfare can also be employed in some cases to deter civilian interference or distract the adversary's attention away from the impending operations.

Lastly, successful completion of special operations can yield valuable Psy warfare themes to further intimidate and/or deter the adversary at the tactical, operational and strategic level as well as to reinforce the will of own population. This was seen after the successful special force operations by India against terrorist camps inside Pakistan's territory in 2017.

Special operations and Psy warfare share a close synergy in the battlefield.

IW and Future Threats

In order to assess the role of IW against future threats, we first need to identify them. While it is hard to predict the future over the long term, several trends are clearly evident and reaching an inflection point over the next decade. Such futuristic threats (and opportunities) involving application of IW are listed below:

- At least one (or more) new country has acquired nuclear weapons every decade starting from 1940s to 2010 and the trend looks to be irreversible. As more and more nations attain nuclear capability, the scope for conventional war keeps shrinking. IW is increasingly being used as the preferred asymmetric option to undermine sovereignty, target individual leaders and engineer social discord. Use of IW to achieve political objectives without resorting to armed conflict will increasingly become the norm in future. Social media platforms like Facebook already face allegations of facilitating regime change through propaganda. ICT firms which control flow of information will thus become important non-state actors with a foothold in the strategic domain and a global footprint. Ability of nation states to survive IW attacks will become a decisive factor in preserving sovereignty and social harmony.

- The number of intelligent machines on the battlefield will increase with advancements in AI and Robotics. Several such unmanned semi-autonomous systems are already operational and we are entering an era of man-machine teams. Many more intelligent weapon systems like drone swarms, combat robots, pilotless aircrafts and intelligent space payloads are under active development. Such machines will eventually pervade all the conventional battle spaces of land, sea, air and space and lead to a paradigm shift in operational strategy, tactics and organizations. These machines will have autonomous computing capability and will communicate, navigate and synchronize using wireless media and electronic sensors. The human element will progressively reduce. Countermeasures and tactics will rely heavily on use of EW and Cyber warfare.
- Space based assets play an increasingly crucial role in surveillance, navigation, communications and targeting. A creeping weaponisation of space is already taking place with anti-satellite weapons and reusable spacecrafts being tested by few countries. Spillover of conventional war into the space domain is thus becoming inevitable. Since outright weaponisation by placing kinetic weapons in orbit can start another arms race; cyber warfare and EW (including DEWs) are going to be a widely used option to deny space based assets.
- The Internet of Things (IoT) is fast becoming a reality with increasingly large number of devices like automobiles, household gadgets, cameras, medical devices and sensors getting networked over the Internet. It is

not only possible to use IoT as a ubiquitous surveillance tool; it can also be used to create a whole new class of cyber attacks spanning the cognitive, physical and electronic domains. Both offensive and defensive cyber warfare comes into play in a major way.

Clearly the role of IW operations is rapidly expanding and many of the emerging threats are ubiquitous in nature and not specific to the mandate of the armed forces. The armed forces as the ultimate defender of national sovereignty however cannot remain aloof. Resolving this conundrum is a huge challenge.

Synergy, Jointness and Integration

Air and naval EW systems are typically integrated with the platforms (ie the aircraft/ship) for purpose of early warning, survival and stealth. Army EW assets on the other hand are mostly platform independent and grouped with field formations. Specialized EW aircrafts and heliborne/UAV based EW assets are a third class of systems which are best suited for integrated application at the theatre level. These can also be integrated for joint operations like amphibious and airborne assaults.

Coordination between land and airborne EW assets is an important aspect requiring joint planning and synergy without which it is possible that air assets target own ground based jammers and vice versa. Processes and procedures would be akin to those which exist between air force and ground based air defense weapons. As regards Cyber and Psy warfare, the capability can be consolidated and unified.

Capability Building

IW capability building can be sub-divided into equipment, training, infrastructure, human resources and leadership. Neither of these can be created overnight. Let us look at these capability building issues one by one.

- **Equipment**. Unlike conventional military weapons and platforms which have a life cycle of between 20 to 30 years, the life cycle of a an EW or Cyber Warfare system is not even ten years because the underlying protocols, software and hardware change at an extremely rapid pace. Take the Windows operating system for example. Ever since computers became a household item we have seen seven generations viz. Windows 95, Windows 98, Windows XP (2001), Windows Vista (2007), Windows 7 (2009), Windows 8 (2012) and Windows 10 (2015) within a span of 20 years. The first iPhone was released in 2007 and the Android operating system was launched around the same time. A whole new ecosystem of smart devices has since come into existence within a decade. Communication, sensors and transmission technologies have similarly evolved at a very rapid pace and weapons to target them must be upgraded every few years. Procurement procedures, financing mechanisms and life cycle concepts have to match with these realities. A separate budget, procurement procedure and policy are therefore warranted in case of IW equipment and systems.

- **Training**. IW domains being highly specialized, personnel need to constantly stay abreast with the latest developments in their domain or else their skills become obsolete in no time. This requires constant updating of skills, specialized hands on training as also continuity on the job particularly in case of the technical disciplines of EW and cyber warfare.
- **Infrastructure**. Considering the extremely specialized skills, a dedicated and integrated IW training infrastructure is a must. The technical disciplines of EW and cyber warfare require hands on training against targets which mimic those present in the actual electronic/ computational terrain. An IW training range comprising of multiple types of electronic transceivers, sensors, computers and networks mimicking the digital battlefield is essential for such training. This is very similar to how field firing ranges are used for training in kinetic weapons.
- **Human Resources Management**. The skill set of an expert hacker or EW specialist is the same as those of a cyber security/ICT expert. Such skilled persons usually get very well paying jobs outside and may not wish to join (or continue) military service. Specific HR policies and incentives are thus required to recruit and retain such highly qualified persons. An alternative model to overcome the problems of training and retaining technical manpower is to outsource some of the IW activities to private contractors provided procedures for same

are in place. The drawback of this model is that serious breaches of classified information similar to what have happened in the USA may happen. Non-military specialists also lack in their understanding of military tactics, terminology and organizations thus limiting their efficiency when it comes to supporting actual operations. A selective combination of the two approaches is possibly the best method.

- **Leadership**. It is impossible to have an individual qualified across all the specializations of psy warfare, intelligence, EW and cyber warfare. In fact it is rare to find an expert who knows everything about even a single discipline, particularly the technical disciplines of EW, SIGINT and cyber warfare. Under these circumstances leadership of IW operations must vest with a specialist who has in depth expertise and hands on experience in at least one core capability with sufficient exposure to at least one technical discipline. Entrusting IW leadership roles to generalists with little clue about technology is a recipe for disaster.

Having acquired a fair understanding of various IW capabilities and their integrated application, it is now time to discern as to how these individual capabilities can be grouped into actual offensive and defensive fighting formation. To do so, we straightaway jump to the two most well known such organizations, namely the US Cyber Command and China's Strategic Support Force. A thorough analysis of these organizations will drive home many practical lessons on how to create force structures for conducting offensive and defensive IW operations in the information battlespace and resolve some of the dilemmas we have outlined earlier.

ENDNOTES

1. Indian Army Doctrine, October 2004
2. Joint Doctrine of Indian Armed Forces, April 2017, Section III
3. Collin T Hunton, "*The War of Ideas and the Role of Information Operations in Counterinsurgency*", April 2007, p47 (School of Advanced Military Studies United States Army Command and General Staff College)
4. Major Patrik Thomé, "*The Role of Information Operations in Strategy, Conventional War and Low Intensity Conflict*", US Air University IO Sphere Journal, Summer 2006

SECTION 3

Doctrine and Force Structures

CHAPTER 6

IW Force Structures of USA

As we have seen in preceding sections – information has emerged as a new battle space and lends itself to both offensive and defensive operations. We have also seen that information being an all pervasive domain; IW can be used to achieve strategic objectives by itself or in conjunction with kinetic means. The former scenario mostly plays out in peacetime and the preparatory stages of war whereas employment in conjunction with kinetic means comes into play once hostilities have broken out and war has commenced. IW-Kinetic integration can also be employed in sub-conventional warfare and counter insurgency operations. It is also clear that offensive and defensive operations in the information battlespace are complementary yet distinct from each other and a balance is required to be maintained between offense and defense.

The two new asymmetric battle spaces of space and information pose a challenge on two fronts – firstly how to integrate these new capabilities into existing operational doctrines and secondly about creation of fighting echelons, headquarters and support elements to optimally exploit these new battle spaces. While theoretically the organizations must flow from doctrine, the situation on ground is often very different and it can be said with confidence that the correctness of any doctrine is best assessed by studying the force structures which have evolved on ground. Let us therefore examine the IW force structures of two modern armies and see how the various capabilities have been operationalized and integrated.

For the purpose of analysis, we have selected the IW force structures of US and Chinese Armed Forces. Reason for selecting them is two fold. Firstly, both USA and China have been at the forefront of ushering in new war fighting paradigms including in the information domain – yet they are also adversarial doctrines arrayed against each other. The moves they make on the strategic chessboard are not only based on rules of the game but also on perceived weaknesses in the opponent's force structure and a long term plan to deliver checkmate. Trying to analyze the underlying logic behind these moves helps us gain many insights. Secondly, both USA and China have restructured and refined their forces to optimally integrate the new and old battlespaces. While

USA which has the maximum IW capabilities has created the unified Cyber Command, China with its stated intent to fight wars under conditions of informatization has responded by creating the PLA Strategic Support Force (SSF) which attempts to integrate both the asymmetric battlespaces of space and information. A study of these organizations helps bridge the gap between theory and actual force structures which can be created and put out to battle.

Structure of the US Armed Forces

The US Armed Forces have a worldwide deployment and are organized into nine unified combatant commands; namely Africa Command, Central Command, European Command, Northern Command, Pacific Command, Southern Command, Special Operations Command, Transportation Command and Strategic Command. The newly created Cyber Command has become the tenth unified command.

These unified commands are collectively spread out over hundreds of large and small military bases/detachments across the globe and there are also forces out at sea and standalone assets in space. Each of these force elements is connected to an information grid which is the lifeblood of the unified operational capability.

As far as the information battlespace is concerned, the US Army, Navy, Air Force and Marine Corps each have their own IW capabilities wielded by specialized units as well as defensive IW elements manning and defending the information grid. Let us look at the US IW capabilities in slightly more detail with an aim to understand how these capabilities have been integrated into the overall force structure. The information has been compiled from publicly available data and may not be absolutely accurate or complete; however it is adequate to allows us to draw some important conclusions, including about the newly raised Cyber Command.

IW Capability of US Army

The US Army has ten Service Component Commands namely the US Army Africa/9th Army, United States Army Central (ARCENT)/Third Army, United States Army Europe (USAREUR)/Seventh Army, United States Army North (ARNORTH)/Fifth Army, United States Army Pacific, United States Army South, United States Army Special Operations Command, Surface Deployment and Distribution Command, United States Army Space and Missile Defense Command and United States Army Cyber Command.

Defensive IW Capability

In 1964 the US Army reorganized its Signals units and activated the US Army Strategic Communications Command to exercise full command and control over worldwide strategic communications, particularly to support the Vietnam War. The name was later changed to US Army Communications Command (USACC). In the mid 1980s digitization and evolution of computer networking led to the USACC being renamed as US Army Information Systems Command (USAISC).

Post operation Desert Storm in the 1990s, a general perception among major commands and theatre commanders was that central management by USAISC deprived them of needed command and control over regional and theatre information systems, computer system acquisitions, and signal assets. The Department of the Army thereafter agreed and dismantled USAISC, relegating the organization to major subordinate command status under US Army Forces Command, and re-designating it as US Army Signal Command (USASC) in 1996.

This decentralization and deregulation led to a proliferation of non-standardized command, control, communications, and computer (C4) systems and an unacceptable level of incompatibility among Army-wide communications equipment and support networks. The department of the Army in the year 2002 once again decided to centralize C4 and information systems management and security under one Army command, namely the Network Enterprise Technology Command (NETCOM)[1]. The 9th Army Signal Command (9th ASC) is the numbered command for NETCOM.

NETCOM's role includes planning, engineering, installing, integrating and protecting the Army's information structure. It also handles the management and defense of the Army frequency spectrum. Formations subordinate to NETCOM[2] are as follows:

- 5th Signal Command (Theater)
  - 2nd Theater Signal Brigade (US Army Europe)

- 7th Signal Command (Theater)
  - 21st Signal Brigade
  - 93rd Signal Brigade
  - 106th Signal Brigade
- 311th Signal Command (Theater)
  - 1st Signal Brigade
  - 516th Signal Brigade
- 335th Signal Command (Theater)
  - 160th Signal Brigade
- Cyber Protection Brigade

NETCOM lies at the heart of US Army's defensive IW capability.

Electronic Warfare Capability

Tactical EW capabilities within the US Army are grouped as integral resources with all Brigade Combat Teams. These include the Ground Auto-Targeting Observation/Reactive Jammer (GATOR), Wolfhound manpack direction finding/location fixing system and the vehicle mounted Prophet mobile SIGINT system[3,4]. Prophet EW systems are also held with the Battlefield Surveillance Brigades. These EW systems collectively facilitate force protection, situational awareness and also integration with kinetic means.

Radio Controlled Improvised Explosive Devices (RCIEDs) have been the main threat to US land forces in Iraq and Afghanistan for past many years. Accordingly, a large number of Counter RCIED Electronic Warfare (CREW) systems like the AN/VLQ-12 Duke and Baldr individual protection equipment are also in service. In fact over 32,000 Duke systems have apparently been fielded on Army combat platforms operating in Operation Enduring Freedom and Operation New Dawn.

Excessive focus on RCIEDs and airborne EW systems to suppress enemy air defenses appears to have led to Army's ground based Integrated EW systems getting phased out and a decline in the overall offensive ground based EW capability – a critical gap which Russia and China correctly identified and forged ahead. This lacuna is now being reversed with planned induction of new integrated EW systems under the Multi-Function Electronic Warfare (MFEW) program[5,6,7,8].

Psychological Warfare Capability

The US Armed Forces uses the term Military Information Support Operations (MISO) instead of Psy Ops. MISO units of the US Army are part of the US Army Special Operations Command (USASOC) which itself comes under the unified Special Operations Command (USSOCOM). The active psy ops units forming part of USASOC are the 4th Military Information Support Group (four battalions) and 8th Military Information Support Group (three battalions), both airborne. Out of the seven MIS battalions five are associated with European, African, Southern, Central and Pacific Commands. One battalion provides audio-visual material support to both the groups while one is a Tactical MISO battalion[9]. There are also two Army reserve formations, namely 2nd and 7th Psychological Operations Groups under the Army Civil Affairs & Psychological Operations Command (Airborne)/USACAPOC[10].

Intelligence and Cyber Warfare Capability

The United States Army Intelligence and Security Command (INSCOM) is a direct reporting unit that conducts intelligence and security operations for US Army commanders and national decision makers. INSCOM collects operational intelligence in all intelligence disciplines including HUMINT. It also conducts production activities like intelligence preparation of the battlefield, SIGINT, Imagery Intelligence (IMINT) and Technical Intelligence (TECHINT) analysis. INSCOM also has responsibilities in counter-intelligence, force protection, electronic and cyber warfare domains.

There are several Military Intelligence (MI) Groups/Brigades under INSCOM. Some of these are grouped with theatre commands viz 66 Military Intelligence Brigade (European Command), 207 Military Intelligence Brigade (Africa Command), 470 Military Intelligence Brigade (Southern Command), 500 Military Intelligence Brigade (Pacific Command), 501 Military Intelligence Brigade (Korea), 513 Military Intelligence Brigade (Central Command) while others handle specialized tasks like counter intelligence support (902 Military Intelligence Brigade), aerial intelligence (116 Military Intelligence Brigade), linguistic support (300 Military Intelligence Brigade) etc. INSCOM also provides manpower and resources to the National Security Agency (NSA) for SIGINT role where they are collectively known as Central Security Service (CSS). INSCOM also provides the National Ground Intelligence Centre which produces TECHINT and intelligence products[11].

INSCOM has substantial IW resources in Cyber Warfare, SIGINT and cryptologic disciplines as outlined below:

- **Specialized MI Brigades**
  - *704 Military Intelligence Brigade.* The 704 Military Intelligence Brigade (three battalions) operates under the NSA for full-spectrum SIGINT, computer network and information assurance operations[12].
  - *706 Military Intelligence Group.* The 706 Military Intelligence Group (having one battalion) provides personnel, intelligence assets and technical support to conduct SIGINT operations within the NSA/CSS[13].
  - *780 Military Intelligence Brigade.* The 780 Military Intelligence Brigade is responsible for conduct of SIGINT and defensive cyberspace operations[14, 15]. It has two subordinate Battalions as under:
    - The 781st Military Intelligence Battalion (Vanguard), earlier called the Army Network Warfare Battalion
    - The 782nd Military Intelligence Battalion (Cyber Legion) which works in collaboration with the 706th Military Intelligence Group which in turn supports the CSS
- **1st Information Operations (IO) Command.** The 1st Information Operations (IO) Command of INSCOM is the Army's only active component organization whose mission is to provide Information and Cyberspace operations support to the Army and other Military Forces. A brigade-level organization consisting of two battalions; the 1st IO Command deploys a variety of mission-tailored teams, each configured to meet the supported command's requirements. The IO Command also operates the Army Information Operations Center (AIOC) which extends planning support, intelligence analysis, and technical assistance to deployed IO teams and other military units, agencies, and departments requesting support. 1st IO Command also runs resident IO related courses and deploys mobile training teams worldwide. Training courses include IO Capabilities, Applications and Planning Course (IOCAP); IO Fundamentals Course (IOFC); Military

Deception Planners Course (MDPC); Electronic Warfare Integration Course (EWIC) and the Military Information Support Operations Integration Course (MISOIC). 1st IO Command's Information Warfare/Cyber Opposing Force (IW/Cyber OPFOR) is tasked with providing realistic, live, virtual, and constructive replicated Nation-State, Violent Extremist, and Criminal Cyber and IW capabilities to Army units during training missions and exercises. The1st IO Command has two subordinate battalions called the 1st and 2nd IO Battalions under it. Both these battalions are capable of providing multiple expeditionary Information Operations (IO) teams in support of deployed forces[16]. Task includes social media analysis and engagement to enable force protection and operational security. The type of teams/capabilities that can be provided by 1st IO Command are as given below and the role appears to be defensive IW operations and facilitating operational integration of IW capabilities:

- *IO Field Support Teams (FST).* FSTs provide supported forces with full spectrum IO subject matter expertise. These deployable FSTs leverage the IO Command's reach back planning and analysis to operationally integrate IO and reinforce forward IO Capabilities.
- *IO Vulnerability Assessment Teams (VAT).* VATs assist units in identifying and resolving vulnerabilities across the full spectrum of information operations, to include Operations Security (OPSEC), Physical Security, COMSEC (phone and email), Social Engineering and Wireless Security, as well as Cyberspace operations.
- *Operations Security (OPSEC) Support Element (OSE).* The OSE trains and prepares appointed OPSEC Program Managers and OPSEC Officers to manage unit OPSEC programs and advise their commanders in all OPSEC areas. This element is responsible for planning, training, conducting assessments, and integrating OPSEC throughout the Army.

Summary

Table below summarizes the discerned IW capability of the US Army:

	Offensive Elements	*Defensive Element*
Electronic Warfare (Tactical)	Equipment is grouped with Brigade Combat Teams	Network Enterprise Technology Command (NETCOM) 780 Military Intelligence Brigade and teams from 1st IO Command (INSCOM)
Cyber Warfare and SIGINT	704 and 706 Military Intelligence Brigades (part of INSCOM). Most of the units affiliated to NSA/CSS	
Psy Warfare	Four MISO/Psy Ops Groups under US Army Special Operations Command (USASOC) which in turn is part of the unified Special Operations Command (USSOCOM). Possibly some teams of 1st IO Command also conduct Psy Ops in the Social Media domain.	
Intelligence (possibly also Deception)	INSCOM and subordinate MI Brigades	

The broad division of responsibility for conduct of defensive and offensive IW Ops and related/supporting capabilities appears to be as follows:

- **Defensive IW Ops.** Network Enterprise Technology Command (NETCOM), essentially the Signals element and elements of 1st IO Command.
- **Offensive IW Ops**
  - *Tactical EW.* Grouped with field formations
  - *Cyber Warfare and SIGINT.* Mostly INSCOM units affiliated to NSA/CSS
- **Psy Warfare.** US Army Special Operations Command (USASOC)
- **Intelligence Capabilities.** Intelligence and Security Command (INSCOM) and the subordinate MI Brigades/Groups

IW Capability of US Navy

There are nine components in the operating forces of the US Navy; namely Pacific Fleet, Fleet Forces Command (formerly Atlantic Fleet), Naval Forces Central Command, Naval Forces Europe, Fleet Cyber Command, Naval Special Warfare Command, Operational Test and Evaluation Force, Military Sealift Command and Navy Reserve.

While all ships have inherent IW capability (particularly integrated EW suites), it is the US Fleet Cyber Command (FCC) which is responsible for Navy's communication networks, offensive and defensive cyberspace operations as well as the space and SIGINT capability. FCC reports directly to the Chief of Naval Operations as an Echelon II command. FCC is organized into 27 active commands, 40 Cyber Mission Force units, and 27 reserve commands deployed around the globe[17]. The US 10th Fleet is the operational arm of Fleet Cyber Command and executes its mission through a task force structure similar to other warfare commanders.

The stated mission of 10th Fleet is to serve as central operational authority for networks, cryptologic/SIGINT, information operations, cyber, electronic warfare, and space capabilities in support of forces afloat and ashore; to direct Navy cyberspace operations globally to deter and defeat aggression and to ensure freedom of action to achieve military objectives in and through cyberspace; to organize and direct Navy cryptologic operations worldwide and support information operations and space planning and operations, as directed; to execute cyber missions as directed; to direct, operate, maintain, secure, and defend the Navy's portion of the Global Information Grid; to deliver integrated cyber, information operations, cryptologic, and space capabilities; to deliver a global Navy cyber common operational picture; to develop, coordinate, assess, and prioritize Navy cyber, cryptologic/SIGINT, space, information operations, and electronic warfare requirements; to assess Navy cyber readiness; to manage man, train, and equip functions associated with Navy Component Commander and Service Cryptologic Commander responsibilities; and to exercise administrative and operational control of assigned forces.

FCC/10th Fleet fulfils its missions through a number of task force based organizations. The main ones amongst these are the Naval Network Warfare Command (NNWC), Naval Cyber Defense Operations Command (NCDOC) and several Navy Information Operations Commands (NIOCs) located worldwide.

FCC also serves as the Navy component command to US Cyber Command, the Navy space component to US Strategic Command, and the Navy's Service Cryptologic Component Commander under the NSA/CSS.

Figure 6.1: Organization of US Navy 10th Fleet and Role of Various Elements

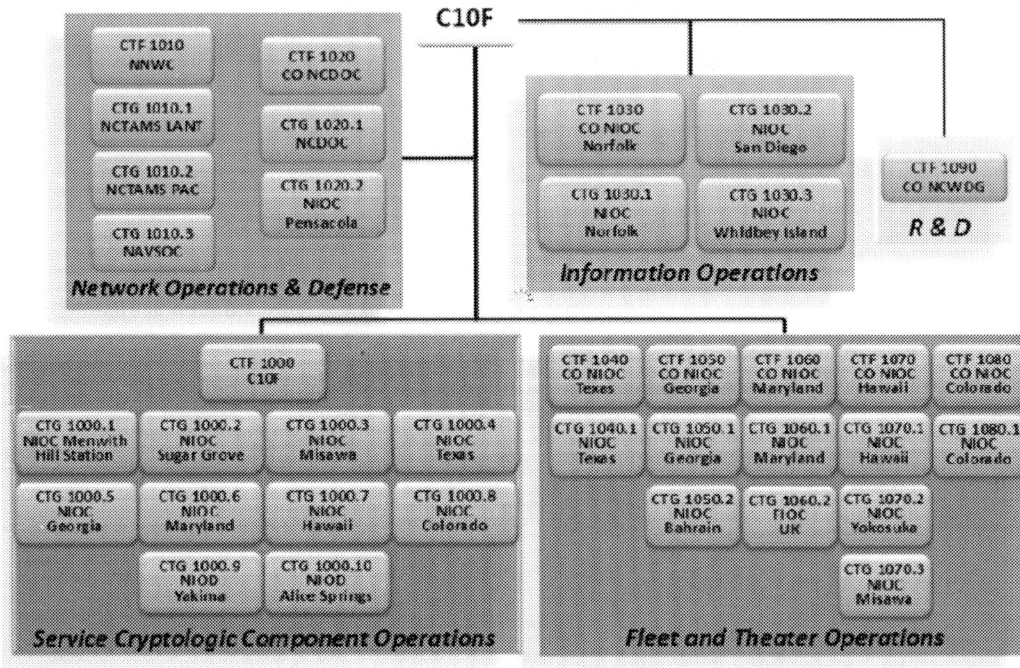

Source: http://www.public.navy.mil/fcc-c10f/Documents/C10F.ORG.pdf

The organization chart below shows the organization of 10th Fleet and the role of the various task forces under it.

Defensive IW Capability

As can be discerned from above diagram, the Navy's defensive IW capability is handled by the NNWC and NCDOC. Brief details of these organizations are as under:

- **Naval Network Warfare Command** (NNWC). In 2002, some 23 organizations from several commands, including the former Naval Space Command, Naval Computer and Telecommunications Command, Fleet Information Warfare Center (FIWC), and Navy Component Task Force – Computer Network Defense (NCTF-CND) were brought together to form the NNWC. With the establishment of US Fleet Cyber Command/US 10th Fleet in 2010, the NNWC was reorganized and its mission revised to operate and defend the

Navy's portion of the Global Information Grid and to deliver reliable, secure Net-centric and Space warfighting capabilities in support of strategic, operational and tactical missions across the Navy. NNWC operates the Naval Computer and Telecommunications Area Master Stations Pacific (NCTAMS PAC) and Atlantic (NCTAMS LANT) as also the Naval Satellite Operations Center (NAVSOC)[18].

- **Navy Cyber Defense Operations Command (NCDOC).** The Naval Computer Incident Response Team (NAVCIRT) was established in 1995 as a department of FIWC. NAVCIRT's primary mission was providing Information Assurance through incident handling and sensor monitoring. A new command, the Navy Component Task Force for Computer Network Defense (NCTF-CND) was likewise created in 1999. The NAVCIRT and NCTF-CND merged in 2003. In 2006, it was renamed as NCDOC and placed under command NNWC. In 2010, NCDOC was placed under the newly created FCC. NCDOC essentially is the Computer Network Defense (CND) service provider for the US Navy[19].

Electronic Warfare Capability

The US Navy has a large number of Electronic Attack/Suppression of Enemy Air Defense (SEAD) squadrons equipped with EA-18G Growler aircrafts. These dedicated EW squadrons are integral to various Carrier Groups (just like ISR capabilities) and not part of FCC/10th Fleet[20]. Individual ships and submarines are obviously equipped with integrated EW suites to jam/evade enemy radars and thwart homing mechanisms of any incoming kinetic projectiles in conjunction with kinetic means. The US Navy also has man pack counter RCIED EW systems like Thor-III[3].

Psychological Warfare Capability

The US Navy does not have any dedicated Psy Ops units except a Naval Reserve PSYOP audiovisual unit[21]. It does however participate in joint Psy Ops by providing assets such as radio broadcast and aerial leaflet dropping as well as provides audio-visual products and imagery. Psy Ops planning and coordination with USAAOC (which has bulk of the Psy Ops units) is carried out by the Naval Network Warfare Command (NETWARCOM) and the Naval Information Operations Command, Norfolk[22] (see below).

Intelligence and Cyber Warfare Capability

The intelligence and cyber warfare capabilities of US Navy are summarized below:

- **Navy Information Operations Commands (NIOCs)**. Several NIOCs exist in the US Navy and are responsible for varied tasks including, SIGINT, EW, cyber, and space related operations in support of the forces. Some of the NIOCs also conduct training in these disciplines including cryptology. Several NIOCs are affiliated to fleets and theatre operations viz. the NIOC at Yokosuka, Japan supports the 7th Fleet of the Pacific Fleet Command. The NIOCs at Colorado, Hawaii, Texas and Georgia likewise provide manpower for the NSA/CSS cryptologic/ SIGINT centres[23].
- **Cryptologic Warfare Group 6**. In 2017, the NIOC at Maryland was re-designated as Cryptologic Warfare Group 6 (CWG-6). In addition to CWG-6, three new commands namely Cryptologic Warfare Maritime Activity (CWMA) 61, Cyber Strike Activity (CSA) 63 and Cyber Defense Activity (CDA) 64 were established to serve as subordinate commands to CWG-6. The Cryptologic Warfare Group appears to be responsible for providing mostly SIGINT (cryptologic) support to the Navy. It also provides manpower to the NSA[24,25,26].
- **Naval Information Forces (NAVIFOR)/Naval Information Dominance Forces (NAVIDFOR)**. The Naval Information Dominance Corps, was created within the US Navy in 2009 to more effectively and collaboratively lead and manage officers, enlisted, and civilian professionals who possess extensive skills in information intensive fields. This corps of professionals works, trains and is educated in information, intelligence, counterintelligence, human-derived information, networks, space, and oceanographic disciplines and tasked with developing and delivering dominant information capabilities in support of US Navy, Joint and national warfighting requirements. In 2016, the command was renamed from Navy Information Dominance Forces (NAVIDFOR) to Naval Information Forces (NAVIFOR) as part of the alignment of the Information Dominance Corps under its new name, the Information Warfare Community. NAVIFOR is an Echelon III command reporting to Commander, US Fleet Forces Command and not FCC. It is the Type Commander (TYCOM) for

meteorology and oceanography, cryptology/SIGINT, cyber, electronic warfare, information operations, intelligence, networks, and space disciplines. Like other TYCOMs – NAVIFOR is primarily responsible for administrative, personnel and training functions. The elements under operational control of FCC are therefore under NAVIFOR for administrative control[27,28].

- **Office of Naval Intelligence (ONI)**. ONI is a core element of the Navy's IW Community which collects, analyzes and produces relevant maritime intelligence and disseminates that intelligence to key strategic, operational, and tactical decision makers to meet Navy, DoD, and national requirements. ONI produces maritime intelligence on naval weapons and technology proliferation, transnational threats in civil maritime counter-proliferation, counter-narcotics and the global maritime environment to strengthen the Navy's conventional and irregular war fighting capacities, and provide foresight into new technologies, future platforms, weapons, sensors, C4ISR and cyber capabilities. ONI has four subordinate analytical centers namely the Nimitz Operational Intelligence Center, Farragut Technical Analysis Center, Kennedy Irregular Warfare Center and Hopper Information Services Center[29].

Summary

Table below summarizes the discerned IW capability of the US Navy:

	Offensive Elements	*Defensive Element*
Electronic Warfare	Electronic Attack/Suppression of Enemy Air Defense (SEAD) squadrons equipped with EA-18G Growler aircrafts which are integral to various Carrier Groups	Naval Network Warfare Command (NNWC), Navy Cyber Defense Operations Command (NCDOC) and Cryptologic Warfare Group 6
Cyber Warfare (including SIGINT)	NIOCs including those affiliated with NSA/CSS and possibly some elements of Cryptologic Warfare Group 6	
Psy Warfare	Mostly provides assets to, and coordinates with US Army Special Operations Command (USASOC) which in turn is part of the unified Special Operations Command (USSOCOM)	
Maritime Intelligence	Office of Naval Intelligence (ONI).	

The broad division of responsibility for conduct of defensive and offensive IW Ops and related/supporting capabilities appears to be as follows:

- **Defensive IW Ops.** Naval Network Warfare Command (NNWC), Navy Cyber Defense Operations Command (NCDOC) and CWG 6 all part of FCC.
- **Offensive IW Ops**
  - *EW.* Integral to carrier groups
  - *Cyber Warfare and SIGINT.* NIOCs including those associated with NSA/CSS, NIOCs supporting various fleets and possibly also CWG 6 with respect to code breaking.
- **Psy Warfare.** Mostly handled by Army Special Operations Command (USASOC). US Navy lends support by providing assets.
- **Intelligence/Related Capabilities.** ONI.

IW Capability of US Air Force

The US Air Force is organized into several major commands like the Air Combat Command, Pacific Air Forces, Global Strike Command, Space Command, Special Operations Command, Air Mobility Command, Material Command, Training Command and the Air National Guard Reserve Force etc. Numbered Air Forces are tactical echelons directly under the major commands. Just like the US Army and Navy, the Air Force also has a large IW capability.

Also the US Air Force prefers to use the terminology "weapon system" for defensive communication/networking related capabilities of Cyber Defense, Cyberspace Defense Analysis, Cyberspace Vulnerability Assessment, Cyberspace Command and Control (C2) Mission System, Cyber Security and Control System and Air Force Intranet Control.

Defensive IW Capability

Each of the numbered Air Force has an Air Operations Center and numerous dedicated Air and Space Communication Squadrons handling terrestrial, space as well as airborne assets to provide dedicated communication and air traffic control support. In addition the US Air Force Space Command (AFSPC) is responsible for worldwide operations through the use of many different types of satellites and launch capabilities. AFSPC also operates the Air Force Network Integration Center and the Air Force Spectrum Management Office.

AFSPC has two subordinate numbered Air Forces namely the 14th and 24th Air Force under command. The 14th Air Force handles capabilities of space lift; position, navigation and timing; satellite communications; missile warning and space control whereas the 24th Air Force (also called Air Forces Cyber) operates, maintains and defends the network capabilities. The AFSPC also controls the Space and Missile Systems Center which oversees launches and supports the Global Positioning, Military Strategic and Tactical Relay (MILSTAR) satellite constellation, expendable launch vehicles, defense meteorological satellites etc.

While the US Air Force liberally and extensively uses the term cyberspace; the AFSPC essentially appears to be a manager of the Air Forces global communication network as well as its space based capabilities. Within AFSPC, the 24th Air Force and its subordinate units (624th Operations Centre, 5th Combat Communications Group, 67th Cyberspace Wing and 688th Cyberspace Wing) appear to be responsible for manning and defending the global communication network as also some multidisciplinary tasks.

A brief breakdown of the units and tasks of the 24th Air Force is as under:

- **5th Combat Communications Group**. The 5th Combat Communications Group (CCG) is an Air Force Space Command unit reporting to 24th Air Force. The group's two combat communications mission squadrons and one support squadron deploy in support of joint task force, combatant command, and Air Force flying wing operations and exercises. The 5th CCG also serves in an advisory capacity for two Air National Guard CCGs and Air Force Reserve Combat Communications squadrons with have a strength of more than 2,500 people located throughout the United States[30,31]. The units subordinate to the 5th CCG are as under:
    - *5th Combat Communications Support Squadron (CBCSS).* The 5th CBCSS houses the capabilities to train and equip 5th CCG's two mission squadrons. The squadron has its own supply system to include an armory capable of fully equipping and deploying communications teams. Additionally, the training flight offers technical communications courses and an Air Education and Training Command certified Combat Readiness School.

- *51st Combat Communications Squadron (CBCS).* The 51st CBCS's primary capability is the large communications package (LCP), designed to support an expeditionary unit of up to 3,000 personnel. Additionally, the 51st CBCS supports the US Secret Service with communications personnel and expertise. The squadron also boasts the Hammer Adaptive Communications Element – a rapidly deployable, highly specialized team capable of responding to incidents within three hours of notification.
- *52nd Combat Communications Squadron.* The 52nd CBCS's medium communications package is a scaled down version of the LCP offering more flexibility in deployment as well as a smaller footprint upon setup. The MCP is capable of supporting smaller deployed units of up to 400 personnel. The squadron also owns the Rapid Access Package, capable of deploying within three hours of notification.

➢ **624 Operations Center.** The 624th Operations Center (OC) serves as the 24th Air Force command and control element and operates the Cyber Command and Control Mission System. The 624th OC consists of four divisions, namely ISR, Combat Plans, Combat Operations and Strategy. The unit reports directly to the 24th Air Force commander. The 624th OC receives orders and tasks from US Cyber Command and in turn tasks 24th Air Force subordinate units to perform a wide range of missions in support of Air Force and joint force commanders[32].

➢ **67th Cyberspace Wing.** Initially activated as the 67th tactical Reconnaissance Wing in the 1940s/50s, it was renamed as the 67th Intelligence Wing in 1993; the 67th Cyberspace Wing (CW) is the Air Force's premier tactical cyberspace warfighting organization dedicated to both offensive and defensive cyber and network operations across the Air Force enterprise and across the interconnected analog and digital portion of the battlespace, comprising of more than 2,500 Airmen, civilians, and contractors across three operations groups with 31 units at 17 worldwide locations. The 67th CW comes under the 24th Air Force. The Wing and its squadrons do not appear to have any aircrafts. Units subordinate to 67th Cyberspace Wing are as under:

- *26th Cyberspace Operations Group.* The 26th Cyberspace Operations Group provides Air Force Network Operations assurance for system and network availability, information delivery, information protection and net-centric enterprise services. The group operates, manages, and secures the network battlespace as part of Air Force Network Operations Command. It employs 3 operational units and over 700 Airmen worldwide, providing battlespace awareness and defense of Air Force Global Information Grid networks and also trains personnel. There are four squadrons under the group as follows:
  - *26th Operations Support Squadron.* The 26 OSS is comprised of six flights. The Intelligence Flight (IN) provides the 67 CW and its subordinate units with threat analysis and up-to-date threat data. The Network Operations Flight provides the 67 CW with financial and project management support. The Weapons and Tactics Flight is the 67 CW's focal point for developing and disseminating tactics, techniques & procedures; and assessing and exercising adversary and friendly tactics and systems. The Current Operations Flight provides operations support to standardize and plan current and future operations, requirements, and capabilities. The flight is also responsible for the 67 CW Coordination Center, providing daily network defensive operations situational awareness and statistical analysis on network defense operations. The Operations Training Flight is the focal point for developing, disseminating, standardizing, and validating training programs across the 67 CW. The Operations Plans Flight supports 67 CW operations by providing dedicated operational-to-tactical level planning; linking 24 AF operations plans to 67 CW execution plans. Essentially it appears to be handling the planning, procurement and training functionalities.
  - *26th Network Operations Squadron.* The 26th NOS delivers persistent Defensive Cyberspace and DoD Information Network (DoDIN) Operations, and Cyber Security Service Provider support to the 24th Air Force via the employment of the Air Force Intranet Control (AFINC) cyber weapon

system and the Joint Regional Security Stack (JRSS) (Air Force Intranet Control). Essentially the squadron appears to be responsible for running the Air Force Intranet.

- *33rd Network Warfare Squadron.* The squadron provides centralized cyberspace defense and mission assurance utilizing the Air Force Cyberspace Defense (ACD) Weapon System across the Air Force classified and unclassified networks. The unit directly supports over one million users, providing cyberspace effects including incident detection and response, digital forensics, and in-depth analysis of network and system logs, and situational awareness by leveraging an expansive fleet of 290 world-wide sensors. The squadron appears to be dedicated to CND/cyber security functions.
- *68th Network Warfare Squadron.* The 68NWS employs mission-ready intelligence professionals who utilize the Cyberspace Defense Analysis (CDA) weapon system to execute Defensive Cyberspace Operations to provide data loss vulnerability assessments, force protection analysis, network vulnerability reporting, and information damage assessments for all Air Force operations, missions, and functions across the AFIN. Once again this is essentially a cyber security unit.

- *67th Cyberspace Operations Group.* The 67 Cyberspace Operations Group provides forces to conduct effective Air Force network attack, network support and communications security of Air Force units, focused on the priorities of the Air Force Space Command Commander, US Strategic Command Commander, and supported Unified Combatant Commanders. The subordinate squadrons[33] are as under:
  - *91st Cyberspace Operations Squadron.* The squadron degrades, disrupts, deceives & exploits adversary telephony and information systems.
  - *352nd Cyberspace Operations Squadron.* The squadron provides operational forces and capabilities to plan and execute cyberspace ISR as well as offensive cyberspace operations.
  - *315th Cyberspace Operations Squadron.* The squadron executes cyberspace attack and cyberspace ISR missions.

- *390th Cyberspace Operations Squadron.* The squadron conducts operations that manipulate the cyberspace domain to enable Combatant Commanders the ability to perform Global Strike and cyber ISR.

- *690th Cyberspace Operations Group.* The group provides anytime/ anywhere support to weapon systems used by network warfighters to execute Network Warfare. The subordinate squadrons are as under:
  - *83rd Network Operations Squadron.* 83rd Network Operations Squadron assures information flow from sensor to the shooter providing cyberspace effects for the Air Force by providing directory and authentication services, monitoring, server storage and backup management, infrastructure, boundary protection, client security, and vulnerability management.
  - *561st Network Operations Squadron.* 561 Network Operations Squadron delivers global connectivity across 109 bases falling under ten major commands and nine combatant commands. The squadron is responsible for the availability and security of both the unclassified and classified secret networks.
  - *690th Network Support Squadron.* The squadron provides the AF Mission Assurance Center (AMAC) for constantly monitoring the AFNet.
  - *690th Cyberspace Operations Squadron.* 690th Cyberspace Operations Squadron located at Hawaii operates, sustains, and defends the Air Force Information Network by performing defensive cyberspace operations employing the cyberspace security and control system.
  - *691st Cyberspace Operations Squadron.* 691st Cyberspace Operations Squadron located at Ramstein AB, Germany. Task is identical to 690th Cyberspace Operations Squadron.

➢ **688th Cyberspace Wing ex 24th Air Force**. The 688th Cyberspace Wing is again part of the 24th Air Force and executes a diverse mission set of cyberspace capability development, test, and assessment; develops and validates cyber tactics; integrates cyber into Air Force Warfare Center training events; employs cyber protection teams to defend priority Department of Defense networks and systems against priority threats; and operates the Air Force cyber and information operations formal training units. The Wing also innovates and partners with government, industry and academia to deliver mission-essential cyberspace capability solutions, and provides cyber range, and modeling and simulation capabilities to Air Force and Joint forces. Additionally, the Wing provides expert and rapid engineering planning, implementation, and installation capabilities to deliver the latest cyberspace infrastructure systems and equipment to mission partners worldwide. The units which form part of 688th Cyberspace Wing[34] are as under:

- *38th Cyberspace Engineering Installation Group.* The 38th Cyberspace Engineering Installation Group (38 CEIG) headquartered at Tinker Air Force Base (AFB), Oklahoma, is the Air Force's premier engineering and installation group. The Group comprises of five squadrons as under:
    - *38th Operations Support Squadron.* Its Mission Support Flight provides an enterprise database that tracks Engineering Installation (EI) requirements from cradle to grave and provides a collaborative environment in which communicators from across the enterprise can jointly manage the cyberspace infrastructure. The flight also manages personnel, manpower, training and security for civilian and military personnel at three different bases. Finally, the flight provides management and oversight of the Group's Non-classified Internet Protocol (NIPR) and Secure Internet Protocol (SIPR) networks including help desk support for the standard desktop configuration and workstation trouble shooting/resolution as an extension of the base Network Control Center. Its Resources Support Flight provides timely and relevant financial management information, financial analysis, and provides for the planning, budgeting, and execution of 38 CEIG mission needs in addition to managing and providing an audit trail for obligation and expenditure of Air Force Space Command.
    - *38th Engineering Squadron.* The primary mission of the 38 ES is to establish, extend and support the cyberspace enterprise

infrastructure by base-lining, planning, designing, establishing, expanding, optimizing and/or replacing cyberspace systems. The 38 ES resources provide expertise to design, engineer, install, configure and test the full range of cyber infrastructure systems, equipment and components typically found at main operating bases as well as other specialty systems unique to expeditionary and strategic communications forces.

- *38th Contracting Squadron.* The Squadron provides contracting support for procuring communications and cyberspace requirements essential to the Air Force's mission though their subordinate flights.
- *38th Cyberspace Readiness Squadron.* The Squadron is responsible for all Air Force and supported Combatant Command long haul communication requirements, provisioning, budgeting and management. This responsibility also includes AF Mobile Satellite System and Teleport services. The Squadron also provides the Air Force a single, consolidated Major Command (MAJCOM) systems management, subject matter expert support and Equipment Control Officers for AF common Communications and Information systems including Personal Wireless Communications Systems (PWCS), Cable/Antenna, IT Asset Management (ITAM) and the Military Auxiliary Radio System.
- *85th Engineering Installation Squadron.* The 85 EIS provides services to mission partners that build and extend the cyberspace enterprise through design, engineering, installation, modification and repair of active components, fiber optics and copper cables. The Squadron also provides support for airfield systems including precision approach systems, meteorological systems and ground radar systems. Additionally, the unit conducts electromagnetic surveys and analysis to assess radio frequency hazards to nuclear weapons, de-conflict interference with Remote Piloted Aircraft and ensure Missile Defense Agency radars work as designed.

- *318th Cyberspace Engineering Installation Group.* Originally raised as the 8th Reconnaissance Group and subsequently reconstituted and re-designated as the 318th EW Group. It was thereafter reactivated as 318th Information Operations Group in year 2000 and renamed as the 318th Cyberspace Operations Group in 2013. The squadrons under this group are as under:

- *39th Information Operations Squadron.* Provides initial qualification training for Airmen to plan and execute missions requiring Cyberspace Operations and Information Operations capabilities.
- *90th Cyberspace Operations Squadron.* 90th COS develops tailored cyber capabilities for cyber defense and cyber-attacks.
- *92nd Cyberspace Operations Squadron.* The 92 COS operates the Cyberspace Vulnerability Assessment/Hunter (CVA/H) software suite for evaluation and classification of security vulnerabilities by detecting deviations from standards, regulation, and penetration testing. It performs Communications Security (COMSEC) assessments and provides Cyber Protection Teams (CPTs) to service and joint Combatant Commanders.
- *318th Operations Support Squadron.* 318 OSS is responsible for development and standardization of operations training, operations synchronization and management of the 688th Cyberspace Wing's Exercise, Intelligence and Weapons and Tactics programs.
- *346th Test Squadron.* 346 TS conducts operational tests, under realistic combat conditions, for any item or component of a weapons system, equipment, or software for the purpose of determining its operational effectiveness & suitability before fielding.

- *688th Cyberspace Operations Group.* The 688th Cyberspace Operations Group (688 COG), plans and executes cyberspace operations to assist a supported commander to fight in and through a contested, friendly cyberspace environment. The 688 COG conducts clear-hold-build operations to remove adversary capabilities if present, defends the supported commander's key cyberspace terrain and critical assets and prepares local cyberspace defenders to sustain advanced cyberspace defense tactics, techniques and procedures to ensure freedom of action within friendly cyberspace while denying adversaries the same. It comprises of the following squadrons:

- *833rd Cyberspace Operations Squadron.* The 833 COS is one of five defensive cyberspace operations squadrons under the 688 COG. The 833 COS presents forces to the Cyber National Mission Force (CNMF) which is responsible for defending National informational networks against cyber threats. It is responsible for organizing, training and equipping three Cyberspace Protection Teams (CPTs).
- *834th Cyberspace Operations Squadron.* 835 COS conducts defensive cyber operations by employing the Cyberspace Vulnerability Assessment/Hunter (CVA/H) software and partners USSTRATCOM and USNORTHCOM.
- *835th Cyberspace Operations Squadron.* 835 COS conducts defensive cyber operations by employing the Cyberspace Vulnerability Assessment/Hunter (CVA/H) software and provides three CPTs.
- *836th Cyberspace Operations Squadron.* Same as 833 COS
- *837th Cyberspace Operations Squadron.* Same as above. It controls three CPTs and partners United States Transportation Command (USTRANSCOM) and Joint Force Headquarters-DOD Network (JFHQ-DoDIN).

As can be seen, bulk of the 24th Air Force is dedicated to providing communications and defending the networks. Only a small number of units, for example the 67th Cyberspace Operations Group (part of 67th Cyberspace Wing) and 90th Cyberspace Operations Squadron (part of 688 Cyberspace Wing) seem to have some offensive role.

Electronic Warfare Capability

Most of the units with Electronic Attack (ECM) capability form part of the Air Combat Command which has four numbered Air Forces namely the 1st, 9th, 12th and 25th numbered Air Forces. Bulk of the Electronic Attack/SEAD capable aircraft like the EC-130H Compass Call are part of the 55th Electronic

Combat Group (three squadrons). The 55th Electronic Combat Group forms part of 55th Wing/25th Air Force (Air Force Intelligence)/Air Combat Command[35]. Thus EW capability can be deemed to be under operational control of the Air Force Intelligence/Air Combat Command.

Psychological Warfare Capability

The 193rd Special Operations Wing which is part of Air National Guard (a reserve force) has squadrons equipped with EC-130J aircraft which can broadcast radio and television signals to target populations from an airborne transmitter, jamming existing television and radio signals where necessary[36]. The US Air Force psy warfare resources appear to be limited and most likely it acts as an asset provider.

Intelligence and Cyber Warfare Capability

Bulk of the intelligence and cyber warfare related capabilities of US Air Force appear concentrated in the 25th Air Force which is part of the Air Combat Command. 25th Air Force is also called Air Force Intelligence and it not only provides multisource intelligence, surveillance, and reconnaissance (ISR) products, applications, capabilities and resources, to include cyber and geospatial forces and expertise, it is additionally also the Service Cryptologic Component responsible to the NSA/CSS for Air Force matters involving the conduct of cryptologic activities, including the full spectrum of missions directly related to both tactical warfighting and national-level operations[37]. As regards AWACS and UCAVs, these are held by Air Combat as well as other commands. Brief details are as under:

- **25th Air Force.** A brief description of the Wings of 25th Air Force is as under:
    - *9th Reconnaissance Wing.* The Wing operates the U-2 Dragon Lady ISR aircraft and the RQ-4 Global Hawk UAV[38].
    - *55th Wing.* This is the largest wing of Air Combat Command and comprises of six groups namely the 55th Communications Group, 55th Electronic Combat Group, 55th Operations Group, 55th Mission Support Group, 55th Maintenance Group and 55th Medical Group. Out of these, the 55th Communications Group provides worldwide C4 and information management. The 55th Electronic Combat Group operates the EC-130H Compass Call

Electronic Attack aircraft while the 55th Operations Group operates the RC-135, OC-135 and WC-135 reconnaissance aircrafts[39].

- *70th ISR Wing*. The wing works closely with the NSA/CSS, leveraging the net-centric capabilities of a worldwide cryptologic enterprise to conduct national missions and enable national-tactical integration for joint and Air Force combat operations around the world. The wing includes six operational intelligence groups located in the US, Pacific and European theatres[40].
- *363rd ISR Wing*. The Wing and its three subordinate Groups have multiple roles primarily focused on precision targeting and ISR analysis for special operations by the Air Force. Task also includes providing direct threat warning and enhanced situational awareness to Air Force Special Operations Command aircrews as well as integrating kinetic and non-kinetic targeting for the Air Force[41].
- *480th ISR Wing*. The Wing comprises six subordinate ISR Groups which conduct real-time tactical and national intelligence collection and dissemination. The Wing also provides cryptologic services[42].
- *319th Air Base Wing*. The 319thAir Base Wing supports the Global Hawk High-Altitude ISR missions and operates the High Frequency Global Communication System[43].

➢ **AWACS**. The E-3 Sentry AWACS aircraft which also perform Command Control and Battle Management (C2BM) functions are operated by the Air Combat Command as also the Pacific Air Force and the Air Force Reserve Command[44, 45].

➢ **UCAVs**. UCAVs like the MQ-9 Reaper and MQ-1 Predator are operated by the Air Combat Command and the Air Force Special Operations Command[46,47,48].

Summary

Table below summarizes the discerned IW capability of the US Air Force:

<table>
<tr><th></th><th>Offensive Elements</th><th>Defensive Element</th></tr>
<tr><td>Electronic Warfare</td><td>EC-130H Compass Call Electronic Attack aircraft are part of the 55th Electronic Combat Group which in turn is part of 55th Wing/25th Air Force/Air Combat Command.</td><td rowspan="2">24th Air Force (part of Air Force Space Command)</td></tr>
<tr><td>Cyber Warfare (including SIGINT)</td><td>Elements of 24th and 25th Air Force affiliated to NSA/CSS</td></tr>
<tr><td>Psy Warfare</td><td colspan="2">193rd Special Operations Wing, part of Air National Guard (a reserve force) equipped with EC-130J aircraft to jam/broadcast TV and Radio signals. Unit works in collaboration with 4th MISO Group (Army Special Operations Command)</td></tr>
<tr><td>Intelligence and Miscellaneous</td><td colspan="2">ISR capabilities with 25th Air Force/Air Combat Command. AWACS held with Air Combat, Pacific and Reserve Commands. UCAVs held by Air Combat Command as well as Air Force Special Operations Command.</td></tr>
</table>

The broad division of responsibility for conduct of defensive and offensive IW Ops and related/supporting capabilities appears to be as follows:

- **Defensive IW Ops.** Air Force Space Command/24th Air Force
- **Offensive IW Ops**
  - *EW.* Air Combat Command/25th Air Force
  - *Cyber Warfare and SIGINT.* Elements associated with NSA/CSS
- **Psy Warfare.** Mostly handled by Army Special Operations Command (USASOC) with Air Force only providing resources to support
- **Intelligence/Related Capabilities.** 25th Air Force handles ISR capabilities. AWACS and UCAVs are held with Air Combat as well as other Commands.

IW Capability of US Marine Corps and Coast Guards

Without going into details, the US Marine Corps Information Systems Management Branch provides defensive IW capabilities for the Marine Corps Enterprise Network (MCEN) and its interfacing components in order to maintain confidentiality, integrity and availability for the network and its data[49]. In 2017 the Marine Corps created two functional organizations namely the Marine Corps Cyberspace Operations Group (MCCOG) and the Marine

Corps Cyberspace Warfare Group (MCCYWG)[50]. Main task of the MCCOG will be to provide communication support to the marine task forces and to operate and defend the MCEN. The MCCYG is an administrative headquarters that organizes, trains, equips, provides administrative support, manages readiness of assigned forces, and recommends certification and presentation of Cyber Mission Force (CMF) Teams to the Cyber Command.

As regards EW – the Marine Corps used to operate the EA-6B Prowler Electronic Warfare aircraft however these have been phased out and offensive EW support is now being provided by the EA-18G aircrafts of Navy and Air Force[51].

The Marine Corps Intelligence Activity (MCIA) is likewise the intelligence support and coordination center for the Marine Corps Intelligence, Surveillance and Reconnaissance Enterprise. MCIA produces all-source intelligence, provides intelligence services, and conducts intelligence and counterintelligence operations. MCIA is comprised of the service intelligence center, CI/HUMINT Support Company and the Marine Cryptologic Support Battalion (MCSB)[52].

The Marine Corps also has a Cryptologic Support Battalion which is under operational control of NSA for SIGINT, Information Assurance and national tactical integration activities[53].

In case of US Coast Guard, it is the office of the Assistant Commandant for Command, Control, Communications, Computers and Information Technology (C4&IT)/CG-6 which designs, develops, deploys, and maintains C4&IT solutions for the entire Coast Guard[54]. In addition the Coast Guard also has an intelligence branch which functions within the department of homeland security and also has a law enforcement role[55].

IW Capability with Defense Intelligence Organizations

The US Department of Defense (DoD) has five dedicated IW/intelligence organizations under command. These are the National Security Agency (NSA), Central Security Service (CSS), Defense Intelligence Agency (DIA), National Geospatial Intelligence Agency and National Reconnaissance Office. Out of these the NSA and CSS are often clubbed together being closely interlinked and under a single commander.

A brief look at the tasks and functions of these intelligence agencies is as under:

- **National Security Agency (NSA) and Central Security Service (CSS).** NSA conducts surveillance of foreign communications, radar and other electronic systems for foreign intelligence/counterintelligence purposes and to support military operations. The CSS is an affiliate organization which provides cryptologic support to NSA and also partners the cryptologic elements of the Armed Forces (the term cryptologic support essentially pertains to SIGINT operations including code breaking). NSA/CSS also handle the information assurance functionalities of the DoD and are tasked with preventing foreign adversaries from gaining access to classified information[56]. Disclosures about NSAs Echelon program reveal that the NSA is involved in mass surveillance of global communications (SIGINT and CNE operations) far beyond military purposes. Likewise the WannaCry ransomware attack of 2017 was revealed to be based on a stolen computer exploit called Eternal Blue developed by NSA. The NSA has thus been proven not only to stockpile zero day vulnerabilities but also weaponize them to launch computer network attacks on a global scale. The Cyber Warfare capability and some portions of the SIGINT capability of all the branches appear to have been consolidated under the NSA/CSS. NSA/CSS do not have any field agents. Both NSA and CSS are headed by the Director NSA who is a four star military general.
- **Defense Intelligence Agency (DIA).** The DIA is a Department of Defense (DoD) combat support agency which produces, analyzes, and disseminates military intelligence information to combat and non-combat military missions. It serves as the primary manager and producer of foreign military intelligence for the Secretary of Defense, the Joint Chiefs of Staff (JCS), and the Unified Combatant Commands. The DIA specializes in collection of HUMINT and MASINT and is a mix of military employees including Army, Navy, Air Force, Marines and defense civilians many of whom are deployed abroad as part of special task forces and in embassies. The Defense Attaché System is also under the Director of DIA. The DIA is headed by a three star general[57].

- **National Geospatial Intelligence Agency (NGA)**. The NGA delivers geospatial intelligence (GEOINT)[58]. Products include maps, charts and satellite imagery for defense and other agencies. The agency is headed by a three star military/defense civilian officer.
- **National Reconnaissance Office (NRO)**. The NRO is the US Government agency in charge of designing, building, launching, and maintaining America's intelligence satellites. Intelligence functions include monitoring proliferation of weapons of mass destruction, tracking and developing military targeting data and bomb damage assessments[59]. It is headed by the Director NRO (a non-military appointment).

We observe that though these are called intelligence organizations, many of them are actually conducting other IW operations as well. This is particularly true in case of NSA/CSS which consolidates cyber warfare and major portions of the SIGINT capability under a single organization.

US Cyber Command

The US Cyber Command was established in 2009-10 as a sub-unified command subordinate to the Strategic Command. There was a dual hat arrangement existing whereby the Director NSA also headed the Cyber Command. In August 2017 it was announced that the Cyber Command will be elevated to become the tenth unified command. In May 2018, the Cyber Command got its own independent commander and the process of its elevation to become the tenth unified command was completed at least in terms of leadership and separation from NSA.

Even prior to the elevation to a unified command, the US Army, Navy, Air Force, Marine Corps each committed various units to the Cyber Command grouped under the aegis of Army Cyber Command, Fleet Cyber Command, Air Forces Cyber and Marine Corps Forces Cyber respectively. The Coast Guard Cyber Command, although subordinate to the Department of Homeland Security, also has a direct support relationship to Cyber Command.

A broad outline of the US Cyber Command can thus be diagrammatically represented as under:

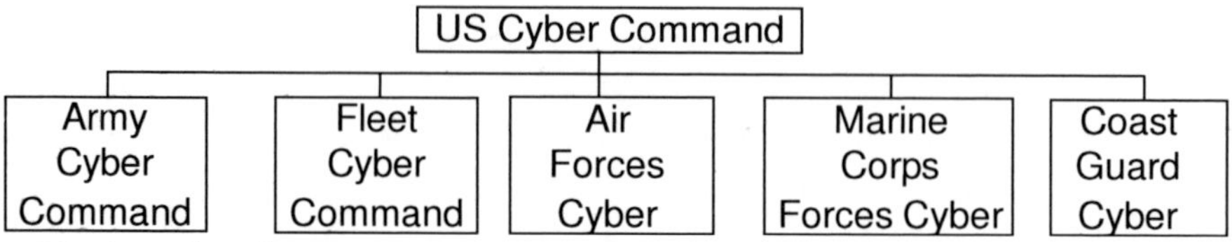

Having already seen the various IW related capabilities and units of the US Army, Navy, Air Force, Marine Corps and Coast Guard in considerable detail; let us examine what all resources have been subordinated by the Army, Navy, Air Force, Marine Corps and Coast Guard towards creation of the Cyber Command.

US Army Cyber Command Elements. The Army has grouped its Network Enterprise Command (NETCOM) and Intelligence and Security Command (INSCOM) under the US Army Cyber Command. The organization is diagrammatically shown below.

Figure 6.2: IW Elements of Army placed under Cyber Command

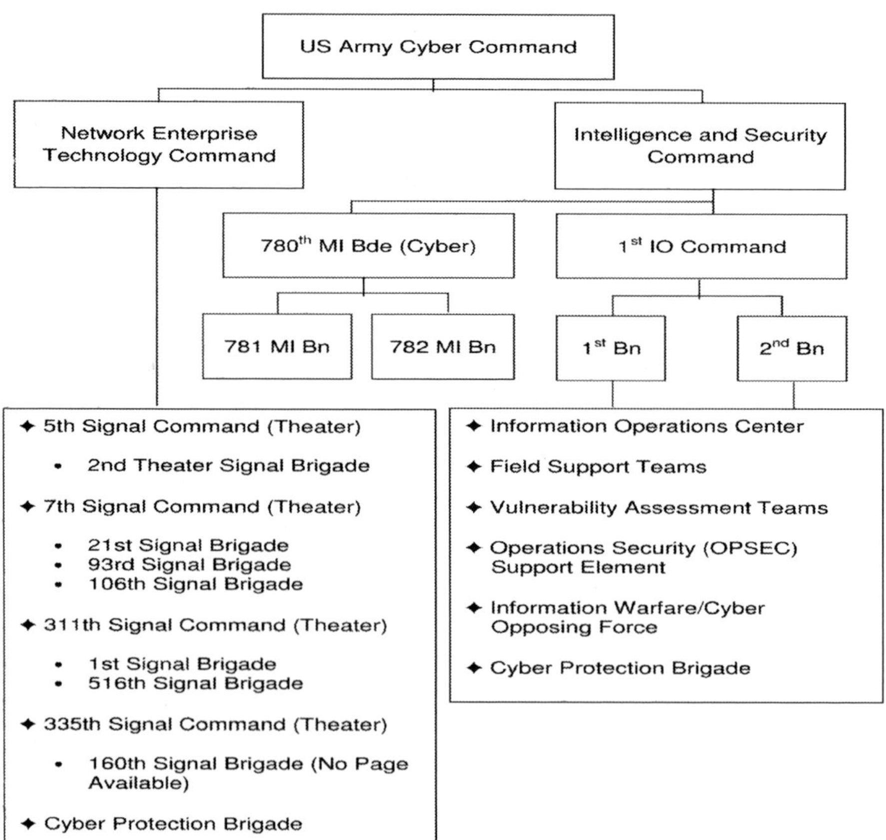

If we see the role and tasking of these units in preceding sections, NETCOMN essentially provides signal communication and networking support to the Army while the 1st IO Command mostly seems to have an OPSEC and COMSEC role. Only the 780th MI Brigade has some SIGINT/ offensive IW functionalities. The 706th MI Brigade (assisting NSA/CSS) has been left out of the ORBAT. The Army portion of Cyber Command is therefore mostly the communication and networking forces and the units responsible for protecting the Army network against attacks. Neither the EW nor the cyber warfare and psy warfare capability seems to be part of Army Cyber Command.

US Navy Cyber Command Elements. The US Navy Fleet Cyber Command has been placed under the US Cyber Command. The organizations are diagrammatically shown below:

Figure 6.3: IW Elements of Navy placed under Cyber Command

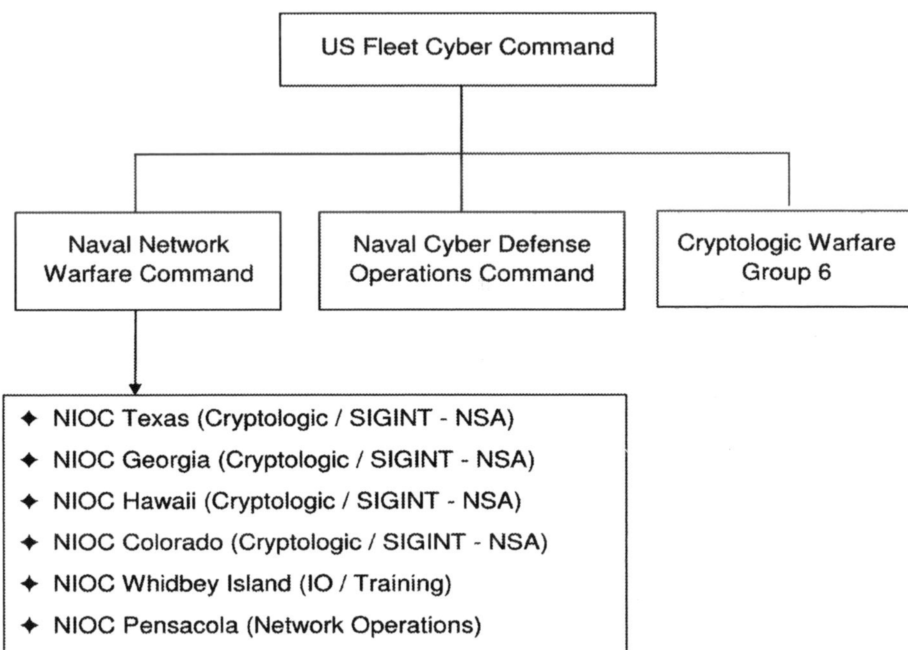

If we examine the role and task of these organizations as covered in the section describing the IW capabilities of US Navy; we find that the Naval Network Warfare Command is the communications/network provider of the Navy, the Naval Information Forces seems to be a training setup for all information related disciplines while the Cryptologic Warfare Group 6 essentially handles encryption related functions. If we leave out the NIOCs working under the NSA/CSS, the US Fleet Cyber Command essentially appears to be a grouping of units providing communications, networking and network security. Neither the EW nor the cyber warfare and psy warfare capability seems to be part of Fleet Cyber Command.

US Air Force Cyber Command Elements. The US Air Force has been a reluctant participant in the Cyber Command venture and has essentially delegated the role to its 24th Air Force. The units comprising the Air Forces Cyber are as depicted below:

Figure 6.4: IW Elements of Air Force placed under Cyber Command

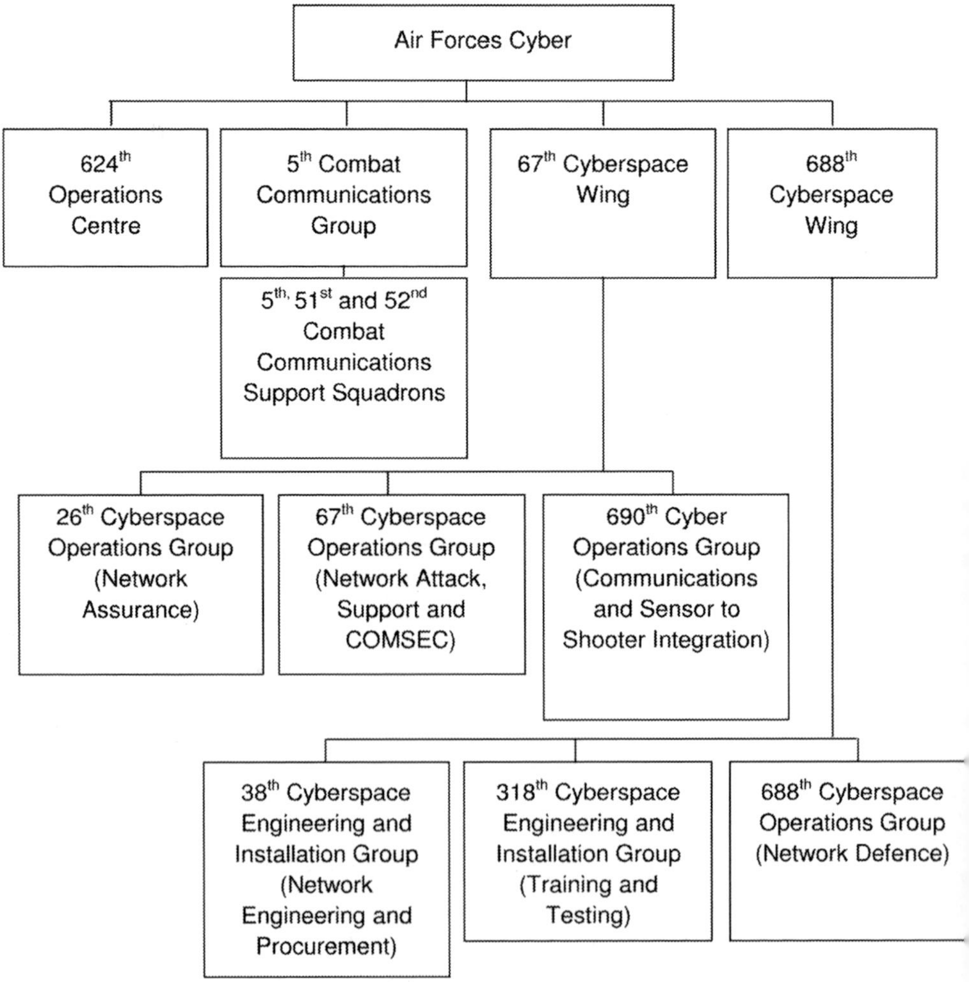

Once again we see that almost all the units of Air Forces Cyber are actually communication, network engineering, training, and procurement units with neither the EW nor the cyber and psy warfare capability being part of Air Forces Cyber.

US Marine Corps and Coast Guard Elements

The Marine Corps Cyberspace Operations Group (MCCOG) and Marine Corps Cyberspace Warfare Group (MCCYWG) have been deputed to be part of the Cyber Command by the Marine Corps. The Assistant Commandant for Command, Control, Communications, Computers and Information Technology (C4&IT)/CG-6 is likewise the Coast Guard element deputed to the Cyber Command. Once again these are purely communication and network defense elements.

IW Capabilities Outside Department of Defense

There are a total of ten intelligence related organizations under the US Department of Defense. Out of these four are the service intelligence components (viz. INSCOM (Army), ONI (Navy), Air Force Intelligence/ 25th Air Force and Marine Corps Intelligence) whereas five are directly under the Department of Defense (viz. DIA, NSA, CSS, NGA and NRO). The Coast Guard Intelligence interfaces with both the Departments of Defense and Homeland Security and can be counted as the tenth one.

Most countries also have a number of civilian intelligence agencies in addition to the military/defense. In case of US, there are seven of them as under:

Sl.No.	*Agency*	*Parent Department*	*Remarks*
1	Central Intelligence Agency	None	Foreign Intelligence
2	Office of Intelligence and Analysis	Homeland Security	Internal intelligence
3	Bureau of Intelligence and Research	State	Intelligence for diplomatic purposes
4	Office of Terrorism and Financial Intelligence	Treasury	Financial intelligence
5	Federal Bureau of Investigation	Justice	Law enforcement and intelligence
6	Office of National Security Intelligence	Justice	Drug Enforcement Administration
7	Office of Intelligence and Counterintelligence	Energy	Protects intellectual property

Intelligence Sharing and Collaboration

Collectively the US has a total of seventeen different organizations and agencies which are in some way involved in gathering intelligence. More than the overlaps, duplication and possible conflicts of interest – it is the aspect of intelligence sharing and interagency cooperation amongst these multiple agencies which is of paramount importance.

The US has tried to optimize this by grouping all the seventeen intelligence agencies as the "US Intelligence Community" through an executive order (Executive Order No 12333 of 1981). The intelligence failure leading to the terrorist attacks of 2001 led to further introspection and enactment of the Intelligence Reform and Terrorism Prevention Act of 2004 as also establishment of the post of a Director of National Intelligence (DNI) and the Office of the Director of National Intelligence (ODNI) to lead the intelligence integration. Several Executive Orders (viz 13354) have tried to strengthen and improve the intelligence sharing[60,61,62].

The Intelligence Community budget is split into the National Intelligence Program and the Military Intelligence Program. Amongst the Director and the Deputy Director of National Intelligence, at least one must be a commissioned officer of the Armed Forces, in active status; or have, by training or experience, an appreciation of military intelligence activities and requirements[63].

ODNI also maintains four national centers that integrate and coordinate the activities of the entire community, or in some cases, broader US Government activities. These are the National Counterterrorism Center (NCTC), the National Counter proliferation Center (NCPC), the National Counterintelligence and Security Center (NCSC), and the Cyber Threat Information Integration Center (CTIIC). Other agencies also have internal mission centers that integrate and coordinate their agency's efforts in a certain mission area, such as CIA's Counterterrorism Mission Center and DIA's regional centers[64].

In addition, several bodies provide oversight and funding to the Intelligence Community to ensure that it works with the law and balances collecting essential information and protecting individuals' interests and privacy. Oversight is exercised by the National Security Council (NSC), President's

Intelligence Advisory Board, The Intelligence Oversight Board, Office of Management and Budget, Privacy & Civil Liberties Oversight Board, Senate Select Committee on Intelligence, House Permanent Select Committee on Intelligence, House and Senate Armed Services Committees, Judiciary Committees, Homeland Security Committees, Appropriations Committees etc. The Intelligence Community also partners various agencies like Foreign Intelligence counterparts of other countries, foreign military counterparts as well as the private sector[65].

From an IW perspective, the most important aspect is the overlapping tasks performed by the Defense Intelligence Agencies (particularly DIA and NSA/CSS) and the CIA. While the CIA has always been the lead agency for HUMINT, it has also been actively involved in Psy Ops. The Vault 7 documents published by Wikileaks indicate that the CIA is also actively involved in cyber warfare and electronic surveillance. One may conclude that the Defense IW apparatus replicates the CIAs role and capabilities (or vice versa) with the only difference possibly being in terms of the targets, the desired intelligence/ operational outcomes, the channel of reporting and degree of covertness. Similar overlaps will exist amongst the other intelligence agencies too because the infosphere in which they all are operating is common and impossible to compartmentalize.

Inferences from the IW Force Structures of USA

A number of conclusions can be drawn from the manner in which the IW capabilities of the US are organized. These are listed below:

- **Defensive IW Operations.** The respective communication branches of the Army, Navy, Air Force, and Marine Corps are responsible for engineering and survivability of global information grid. The Army, Navy, Air Force, and Marine Corps also have dedicated cyber security (CND) units. Together they provide the required information assurance to US forces worldwide. These capabilities have now been placed under the newly created Cyber Command.
- **Electronic Warfare.** Army and Naval EW capability is grouped with fighting formations and carrier groups respectively. Air Force EW resources for SEAD have been consolidated under the Air Combat Command. No EW capability has been placed under the Cyber Command by any of the branches.

- **Psychological Warfare.** MISO groups which are part of Special Operation Command are responsible for conduct of Psy Ops of the Army. The Navy and Air Force provide resources to the Army and are essentially in a supporting role. Possible reason for this is that Army troops are deployed on ground and in direct contact with the target population. Psy warfare capability has again not been placed under the Cyber Command.
- **Cyber Warfare.** Cyber Warfare capability has been consolidated under the NSA/CSS which is an independent intelligence agency directly under the Department of Defense. Major portions of the SIGINT capability (particularly cryptologic) have also been placed under the NSA/CSS highlighting the major synergy between cyber warfare and SIGINT operations. None of the cyber warfare capability has been shed to the newly created Cyber Command.
- **Intelligence Capabilities.** ISR capabilities are grouped with fighting echelons. Operational intelligence relevant to respective branches is handled by dedicated organizations within the Army, Navy, Air Force and Marine Corps. GEOINT and space based intelligence capabilities are handled by the NGA and NRO respectively whereas DIA specializes in collection of HUMINT and MASINT. None of these capabilities have been transferred to the Cyber Command.

What this leads to is the conclusion that the US Cyber Command is essentially a grouping of formations and units which are responsible for providing communications to the US Army, Navy, Air Force, Marine Corps and Coast Guard. It is not designed to launch offensive operations in any of the information terrains. Its purpose instead is to ensure survival of the worldwide communication grid against enemy attacks ie information assurance. Its subordinate units conduct no offensive cyber warfare, ECM, offensive psy ops or intelligence acquisition except what may be required to fulfil defensive tasks.

This should not be surprising because of the expeditionary nature of the US Armed Forces. Since they operate far from their home bases, availability of joint communications and C4ISR capabilities is vital to their survival. Securing own lines of communications is of far greater significance then disrupting the enemy lines of communication. The cyber command has been raised to ensure precisely this. It is the key defensive formation holding ground in the electronic terrain.

The following extract from the history of the US Navy 10th Fleet/Fleet Cyber Command merits attention:

> *"The White House Cyberspace Policy Review of May 2009 stated that "America's failure to protect cyberspace is one of the most urgent national security problems facing the new administration." Two months later, Secretary of Defense Gates unveiled his plan for military cyberspace operations. In a memo to the Secretaries of the Armed Forces, he wrote, "Our increasing dependency on cyberspace, alongside a growing array of cyber threats and vulnerabilities, adds a new element of risk to our national security. To address this risk effectively and to secure freedom of action in cyberspace, the Department of Defense requires a command that possesses the required technical capability and remains focused on the integration of cyberspace operations. Further, this command must be capable of synchronizing war-fighting effects across the global security environment as well as providing support to civil authorities and international partners*[66]*."*

The logic underlying the US Cyber Command thus becomes obvious. The US Cyber Command by design is a defensive holding formation in the electronic type terrain of the information battlespace and it is meant to ensure that the US global information grid is controlled, secured and defended against attacks by the adversary under all circumstances.

ENDNOTES

1. Breslin Vince, "*Network Enterprise Command Evolved from Strategic Communications Command*", (Army Communicator Magazine, US Army School of Signals, June 2010)
2. Website of the US Army Network Enterprise Technology Command (NETCOM), "*Organization of NETCOM*", Web URL: http://netcom.army.mil/about/Organization.aspx accessed on 10 January 2018
3. Website of the Program Executive Office Integrated Electronic Warfare and Systems (PEO IEW&S), "*PM Electronic Warfare and Cyber*" Web URL: https://peoiews.army.mil/programs/pm-ewc/ accessed on 10 January 2018
4. Military and Aerospace Electronics website, "*General Dynamics to upgrade Army AN/MLQ-44A Prophet-Enhanced SIGINT vetronics systems*" (Published on 29 June 2016), Web URL: http://www.militaryaerospace.com/articles/2016/06/sigint-vetronics-prophet.html accessed on 10 January 2018
5. McLeary Paul, "*Russia's Winning the Electronic War*", (Published in Foreign Policy Magazine website, 21 October 2015), Web URL: https://foreignpolicy.com/2015/10/21/russia-winning-the-electronic-war/ accessed on 10 January 2018

6. Col Laurie Moe Buckhout (Retd), "*Short History of US Army Electronic Warfare*", (SITREP Quarterly Newsletter by Leonardo DRS, Q1-2016)
7. US Defense Science Board Study, "*21st Century Military Operations in a Complex Electromagnetic Environment*", (Office of the Under Secretary of Defense for Acquisition, Technology, and Logistics, July 2015), p.6
8. United States Government of Accountability Office Report GAO 12-479 (July 2012) to the Committee on Armed Services, House of Representatives, "*Electronic Warfare : DOD Actions Needed to Strengthen Management and Oversight*", p.2
9. American Special Ops website, "*Psychological Warfare*", Web URL: http://www.americanspecialops.com/psyops/ accessed on 01 May 2018
10. Website of the US Army Reserve, "*Civil Affairs and Psychological Operations Command (Airborne)*", Web URL: http://www.usar.army.mil/USACAPOC/ accessed on 01 May 2018
11. Website of the US Army Intelligence and Security Command (INSCOM), "*Major Subordinate Commands*", Web URL: http://www.inscom.army.mil/MSC/ accessed on 10 January 2018
12. Website of the US Army Intelligence and Security Command (INSCOM), "*704th Military Intelligence Group*", Web URL: https://www.inscom.army.mil/MSC/704MIG.aspx accessed on 10 January 2018
13. Website of the US Army Intelligence and Security Command (INSCOM) "*706th Military Intelligence Group*", Web URL: https://www.inscom.army.mil/MSC/706MIG.aspx accessed on 10 January 2018
14. George I. Seffers, "*Historic Cyber Unit Begins Daily Action*", (Published on the The Armed Forces Communications and Electronics Association website in March 2012), Web URL: https://www.afcea.org/content/?q=node/2891 accessed on 10 January 2018
15. Website of the US Army Intelligence and Security Command (INSCOM), "*780 Military Intelligence Brigade*", Web URL: https://www.inscom.army.mil/MSC/780MIB/units.html accessed on 10 January 2018
16. Website of the US Army Cyber Command, "*1st IO Command*", Web URL: http://www.arcyber.army.mil/Organization/1st-IO-Command/ accessed on 10 January 2018
17. Website of the US Fleet Cyber Command/US Tenth Fleet, Web URL: http://www.public.navy.mil/fcc-c10f/Pages/home.aspx accessed on 01 May 2018
18. Website of the US Fleet Forces Command, "*Naval Network Warfare Command: Our History*", Web URL: http://www.public.navy.mil/FLTFOR/nnwc/Pages/Command-History.aspx accessed on 01 May 2018
19. Website of the US Fleet Forces Command, "*Navy Cyber Defense Operations Command (NCDOC): History*", Web URL: https://www.public.navy.mil/FLTFOR/ncdoc/Pages/History.aspx accessed on 01 May 2018
20. Troy Prince, "*EA-18G Growler Permanent Bases*", Web URL: http://www.vaq136.com/ea18gbases/index.html accessed on 01 May 2018
21. Special Operations Forces Reference Manual, "*US Naval Special Operations Forces*", Web URL: "https://fas.org/irp/agency/dod/socom/sof-ref-2-1/SOFREF_Ch4.htm accessed on 01 May 2018
22. Infogalactic website, "*Psychological Operations (United States)*", Web URL: https://infogalactic.com/info/Psychological_Operations_(United_States)#Navy accessed on 01 May 2018
23. Website of the US National Security Agency/Central Security Service, "*NSA Cryptologic Centers*", Web URL: https://www.nsa.gov/about/cryptologic-centers/ accessed on 01 May 2018

24. Website of the US Fleet Forces Command, "*Cryptologic Warfare Group SIX: History*", Web URL: https://www.public.navy.mil/FLTFOR/cwg6/Pages/History.aspx accessed on 01 May 2018
25. CWG-6 Public Affairs article, "*NIOC Maryland holds Change of Command, Becomes CWG-6*", Web URL: http://www.navy.mil/submit/display.asp?story_id=100972 accessed on 01 May 2018
26. US Department of Navy OPNAV Notice 5400.2237 of 05 Jun 2017
27. US Naval Academy website, "*Information Dominance Corps*", Web URL: www.usna.edu/Cyber/_files/documents/idc/IDC_Overview.pdf accessed on 01 May 2018
28. Website of US Naval Information Forces (NAVIFOR), "*External Links: ADCON Commands*", Web URL: https://www.public.navy.mil/FLTFOR/navifor/Pages/Links.aspx accessed on 01 May 2018
29. Website of the US Office of Naval Intelligence (ONI), "*ONI Fact Sheet*" Web URL: http://www.oni.navy.mil/Portals/12/This_is_ONI/Who%20we%20are/factsheets/ONI_Factsheet_2017.pdf?ver=2017-12-06-080639-193 accessed on 01 May 2018
30. Website of the Robins US Air Force Base, "*5th Combat Communications Group*", Web URL: http://www.robins.af.mil/About-Us/Fact-Sheets/Display/Article/377488/5th-combat-communications-group/ accessed on 01 May 2018
31. Website of US Air Forces Cyber, "*5th Combat Communications Group Fact Sheet*", Web URL: http://www.afcyber.af.mil/About-Us/Fact-Sheets/Display/Article/458564/5th-combat-communications-group/ accessed on 01 May 2018
32. Website of US Air Forces Cyber, "*624th Operations Centre*", Web URL: http://www.afcyber.af.mil/About-Us/Units/624th-Operations-Center/ accessed on 01 May 2018
33. Website of US Air Forces Cyber, "*67th Cyberspace Wing Fact Sheet*", Web URL: http://www.afcyber.af.mil/About-Us/Fact-Sheets/Display/Article/458563/67th-cyberspace-wing/ accessed on 01 May 2018
34. Website of US Air Forces Cyber, "*688th Cyberspace Wing Fact Sheet*", Web URL: http://www.afcyber.af.mil/About-Us/Fact-Sheets/Display/Article/458566/688th-cyberspace-wing/ accessed on 01 May 2018
35. Website of the US Air Force, "*Compass Call Fact Sheet*", Web URL: http://www.af.mil/About-Us/Fact-Sheets/Display/Article/104550/ec-130h-compass-call/ accessed on 01 May 2018
36. Website of the US 193rd Special Operations Wing, "*193rd Special Operations Wing Fact Sheets*", Web URL: http://www.193sow.ang.af.mil/About-Us/Fact-Sheets/Display/Article/439262/193rd-special-operations-wing/ accessed on 01 May 2018
37. Website of the US 25th Air Force, "*US 25th Air Force Fact Sheet*", Web URL: http://www.25af.af.mil/About-Us/Fact-Sheets/Display/Article/662963/twenty-fifth-air-force/ accessed on 01 May 2018
38. Website of the Beale US Air Force Base, "*9th Reconnaissance Wing Fact Sheet*", Web URL: http://www.beale.af.mil/Library/Fact-Sheets/Display/Article/279932/9th-reconnaissance-wing/ accessed on 01 May 2018
39. Website of the Offut US Air Force Base, "*55th Wing Fact Sheet*", Web URL: http://www.offutt.af.mil/Portals/97/55th%20Wing%20Fact%20Sheet2.pdf?ver=2017-08-21-102734-797 accessed on 01 May 2018
40. Website of the US 25th Air Force, "*70th ISR Wing Fact Sheet*", Web URL: http://www.25af.af.mil/About-Us/Fact-Sheets/Display/Article/671802/70th-intelligence-surveillance-and-reconnaissance-wing/ accessed on 01 May 2018
41. Website of the US 25th Air Force, "*363rd ISR Wing Fact Sheet*", Web URL: http://

www.25af.af.mil/About-Us/Fact-Sheets/Display/Article/662965/363rd-isr-wing/ accessed on 01 May 2018

42. Website of the US 25th Air Force, "*480th ISR Wing Fact Sheet*", Web URL: http:// www.25af.af.mil/About-Us/Fact-Sheets/Display/Article/662961/480th-intelligence-surveillance-and-reconnaissance-wing/ accessed on 01 May 2018
43. Website of Grand Forks US Air Force Base, "*319th Air Base Wing*", Web URL: http:// www.grandforks.af.mil/Units/ accessed on 01 May 2018
44. Website of the 522nd Air Control Wing, "*522nd Air Control Wing Fact Sheet*", Web URL: http://www.552acw.acc.af.mil/Library/Fact-Sheets/ accessed on 01 May 2018
45. Senior Airman John Linzmeier, "*Eyes in the Sky: 961st Airborne Air Control Squadron Provides Command, Control Support*", (Published on the US Pacific Command website on 17 April 2017), Web URL: http://www.pacom.mil/Media/News/News-Article-View/Article/1154254/ eyes-in-the-sky-961st-airborne-air-control-squadron-provides-command-control-su/ accessed on 01 May 2018
46. Website of the Beale US Air Force Base, "*348th Reconnaissance Squadron Fact Sheet*", Web URL: http://www.beale.af.mil/Library/Fact-Sheets/Display/Article/774431/348th-reconnaissance-squadron/ accessed on 01 May 2018
47. Website of the Creech US Air Force Base, "*History of Creech Air Force Base*", Web URL: http:/ /www.creech.af.mil/About-Us/Fact-Sheets/Display/Article/449127/history-of-creech-air-force-base/ accessed on 01 May 2018
48. Website of the Cannon US Air Force Base, "*27th Special Operations Group*", Web URL: http:/ /www.cannon.af.mil/Units/27th-Special-Operations-Group/ accessed on 01 May 2018
49. Website of the US Marine Corps, "*Information Systems Management Branch: Cyber Security*" Web URL: http://www.hqmc.marines.mil/ar/Branches/Information-Systems-Management/ Cyber-Security/ accessed on 01 May 2018
50. Marine Corps Concepts and Programs website, "*US Marine Corps Forces, Cyberspace Command (MARFORCYBER)*", Web URL: https://marinecorpsconceptsandprograms.com/organizations/ operating-forces/us-marine-corps-forces-cyberspace-command-marforcyber accessed on 10 January 2018
51. Stars and Stripes website, "*Final Flight: Remembering the Navy's EA-6B Prowler*", (Published on 03 July 2015), Web URL: https://www.stripes.com/news/final-flight-remembering-the-navy-s-ea-6b-prowler-1.356060 accessed on 01 May 2018
52. Website of the US Marine Corps, "*Marine Corps Intelligence Activity*", Web URL: http:// www.hqmc.marines.mil/intelligence/Units/MCIA/ accessed on 01 May 2018
53. Website of the US Marine Corps, "*Marine Cryptologic Support Battalion*", Web URL: http:// www.hqmc.marines.mil/intelligence/Units/Marine-Cryptologic-Support-Battalion/ accessed on 01 May 2018
54. Website of the US Coast Guard, "*Assistant Commandant for C4&IT*", Web URL: https:// www.dcms.uscg.mil/Our-Organization/Assistant-Commandant-for-C4IT-CG-6/ accessed on 01 May 2018
55. Website of the US Intelligence Community, "*US Coast Guard Intelligence*", Web URL: https:/ /www.intelligence.gov/how-the-ic-works/our-organizations/421-us-coast-guard-intelligence accessed on 01 May 2018
56. Website of the US National Security Agency/Central Security Service, "*What We Do*", Web URL: https://www.nsa.gov/what-we-do/ accessed on 01 May 2018
57. Website of the US Defense Intelligence Agency, "*History*", Web URL: http://www.dia.mil/ About/History/ accessed on 01 May 2018

58. Website of the US National Geospatial Intelligence Agency, "*Mission*", Web URL: https://www.nga.mil/ProductsServices/Pages/default.aspx accessed on 01 May 2018
59. Website of the US National Reconnaissance Office, "*About the NRO*", Web URL: http://www.nro.gov/about/index.html accessed on 20 December 2017
60. US Government Executive Order Executive Order 12333 of 04 December 1981 available at Web URL: https://en.wikisource.org/wiki/Executive_Order_12333 accessed on 01 May 2018
61. Cornell Law School, Legal Information Institute website, "*Deputy Directors of National Intelligence*", Web URL: https://www.law.cornell.edu/uscode/text/50/3026 accessed on 01 May 2018
62. US Government Executive Order Executive Order 13354 of 27 August 2004, available at Web URL: https://en.wikisource.org/wiki/Executive_Order_13354 accessed on 01 May 2018
63. Website of the US Director of National Intelligence, "*US Intelligence Community Budget*", Web URL: https://www.dni.gov/index.php/what-we-do/ic-budget accessed on 01 May 2018
64. Website of the US Director of National Intelligence, "*How We Work*", Web URL: https://www.dni.gov/index.php/how-we-work accessed on 01 May 2018
65. Website of the US Intelligence Community, "*How the IC Works*", Web URL: https://www.intelligence.gov/how-the-ic-works#our-organizations accessed on 01 May 2018
66. Website of the US Fleet Cyber Command/US Tenth Fleet, "*History*", Web URL: http://www.public.navy.mil/fcc-c10f/Pages/history3.aspx accessed on 01 May 2018

CHAPTER 7

IW Force Structures of China

The Chinese warfighting doctrine has evolved from "people's war" and "active defense" to "winning local wars under conditions of informatization" and more recently to "winning informationized wars"[1]. This progressive evolution reflects the growing importance of NCW and IW as key elements of Chinese warfighting strategy.

China sees USA as its main challenger and also realizes that the US not only possesses superior weapons technology but also dominates the new battlespaces of space and information. The global information grid of the US military enables its expeditionary forces to project force across the globe and fight a network-centric war. The only way for China to significantly diminish US capability is by crippling this grid through EW, cyber warfare and anti-satellite capability. Cyber warfare can also serve as a means to hasten China's modernization through espionage and technology exfiltration. How much this espionage narrative is true is a matter of conjecture.

The two prongs of China's informatization plan thus become evident – the first is to transform the People's Liberation Army (PLA) into a modern high tech force capable of waging network-centric (informatized) war, and the second one is to develop offensive IW capability to deny similar network-centricity to its adversaries. In adherence to this strategy, China is not only trying to close the gap with the US in high tech systems like aircrafts, ships and missiles but also paying special attention to the new battlespaces of space and information so that it can cripple the US global information grid. China is also investing heavily in futuristic technologies like AI, quantum computing and robotics which are nascent fields with significant military potential and can help China leapfrog ahead of the USA[2]. Since not much information is available in the public domain, a generic overview of Chinese IW capability is given below which is however sufficient to considerably enhance our knowledge of offensive and defensive IW force structures which manifest practically in the information battlespace.

Overview of China's IW Capability

Cyber Warfare

China's growing Cyber Warfare capability is well known. PLA Unit 61398 was identified in 2013 by a US cybersecurity firm called Mandiant (subsequently acquired by FireEye) which dubbed it as an Advanced Persistent Threat (APT) focused on stealing confidential business information and intellectual property. Another entity called Unit 61486 was subsequently identified as yet another APT. Unit 61486 in addition to cyber espionage has also been reported to target space and satellite based assets[3]. Overall FireEye has identified numerous such APTs (essentially state sponsored large hacking groups) operating globally and majority of them are from China[4]. The identity of Unit 61398 has been revealed to be the 2nd Bureau of the 3rd Department of PLA GSD (details covered below).

The massive GhostNet cyber spying operation discovered in March 2009 is likewise an example of Chinese CNE capability. Computer systems belonging to embassies, foreign ministries and other government offices in 103 countries including the Dalai Lama's Tibetan exile centers in India, London and New York City were compromised by GhostNet[5]. Operation Shady RAT is yet another ongoing series of cyber-attacks starting in mid-2006 and making use of Remote Access Tools. It is again widely assumed to be orchestrated by the People's Republic of China and hit at least 71 organizations, including defense contractors, businesses, the United Nations, and the International Olympic Committee[6]. The US Government has repeatedly blamed Chinese hackers for carrying out industrial espionage and theft of blueprints, proprietary manufacturing processes, test results, business plans, pricing documents, cooperation agreements and trade secrets on an unprecedented scale.

It is however pertinent to mention that reports published by US and its allies conveniently omit referring to the US NSA/CSS and CIA as APTs which they indeed are. This too is a subtle form of cognitive IW.

China's Cyber Warfare Doctrine is designed to achieve global "electronic dominance" by 2050 which would include the capability to disrupt the critical infrastructure of their enemies prior to initiation of kinetic operations[7].

In its January 2017 newsletter the Indian National Critical Information Infrastructure Protection Centre (NCIIPC) reported the share of various kinds of attacks on critical infrastructure in year 2016. The report based on total event count of around 7.5 million incidents, threat feeds and vulnerabilities states that China has been the major source of malicious traffic contributing about 70 percent of the total[8].

Electronic Warfare

The PLA is making a concerted effort to develop indigenous EW equipment and systems. China recently unveiled the H-6G electronic warfare aircraft which has been under development since 2008. The aircraft has high electronic jamming power and can cover relatively bigger combat areas such as the South China Sea and East China Sea as per reports. This will be a major boost to its offensive capabilities[9].

In 2017 at a military parade to mark the 90th anniversary of the PLA, China displayed a group of drones that can act as a swarm to electronically paralyze and suppress the enemy's early warning and command/control communications. Global Times quoting Wu Yafei, head of the 'electronic confrontation' formation at the parade stated that among the many armaments that were on parade, there were 16 items of the PLA's latest electronic warfare equipment that can disrupt enemy radar and communication in air defense and field battles[10].

Another key area of focus of Chinese EW planners has been on offensive EW (ECM) to cripple critical space assets of its adversaries located in medium, geosynchronous and eccentric earth orbit where other technologies are less effective. The most important targets are the tactical communications platforms in geosynchronous orbit and the Global Positioning System (GPS) constellation in medium Earth orbit. China's emphasis on electronic attack in the case of such platforms is not simply a function of their orbital altitude, because many of China's current direct-attack kinetic systems could reach distant orbits without great difficulty if dedicated to that purpose. It is more likely that cost and mission effectiveness as well as political considerations have driven Chinese planners towards electronic attack methods in such instances. China's current research, development and acquisition programs therefore seek to neutralize the space-based tactical communications and global positioning constellations not by physical attacks but by a 'denial of service' approach. This has the advantage of avoiding larger conflicts with the international community while promising to impede the adversary military's ability to communicate tactically and secure position, navigation and timing information. China has relentlessly focused on acquiring sophisticated jamming technologies that would permit it to enforce information blackouts at critical moments. Concerns have been raised in the US that EW is an area where Russia and China have possibly surpassed US capability already.

As part of a larger effort to develop 'new concept weapons', China has also devoted considerable resources to directed energy weapons, particularly ground based high and low energy lasers. Other technologies that have been discussed in China include radio frequency weapons, high-power microwave weapons, electromagnetic rail guns and particle-beam systems. Internet and media reports describe China firing a high-powered laser at a US satellite, resulting in it being temporarily blinded in year 2006 and China having destroyed the control chip of a Japanese spy satellite using an electromagnetic pulse weapon nicknamed "Poacher One" in 2014. These reports point to a significant new capability that China continues to develop and improve upon.

Psychological Warfare

Chinese philosophy lays great emphasis on winning without resorting to armed conflict and psy warfare is a cornerstone of this strategy. The Chinese have tried to consolidate the ambit of IW in the cognitive realm in what is popularly referred to as the concept of "Three Warfares".

The "Three Warfares" concept was approved by the Chinese Communist Party (CCP) Central Committee and the Central Military Commission (CMC) in 2003. The constituents of three warfares are "Psychological Warfare" to deter and demoralize the enemy personnel and civilian populations, "Public Opinion or Media Warfare" to influence domestic and international public opinion and build support for China's military actions and lastly "Legal Warfare" to exploit international and domestic law to create a legal high ground for asserting Chinese interests and hamstring the enemy.

Collectively, the three warfares are aimed at preconditioning key areas of competition in China's favor. The concept has been extensively used by China to shape the operational space in South China Sea over the past decade. Recently China also used this concept against India during the 2017 Doklam standoff.

Communications and Networking

The PLA has been able to make improvements in its C4ISR capabilities with the introduction of satellite communications, wireless networks, digital radios and UAVs. The ground forces have been regularly asked to operate under complex ECM conditions in exercises. Furthermore, the PLA is developing a network-centric warfare capability by connecting different combat, intelligence, surveillance and reconnaissance elements to form an integrated network. The term used in Chinese literature is "informationization"[11].

China is also known to have a so called "Blue Army" which performs the role of opposing forces (OPFOR) during exercises. A Chinese Ministry of Defense spokesperson pointed out in 2011 that the PLA had set up an "Internet Blue Army" that specifically dealt with cyber attacks and that it was not into hacking, rather its purpose was to increase the military's own protection levels[12].

Internet, Telecommunications and Media

The IW capability of China is deeply influenced by China's unique single party political system and the resultant military-cum-political nature of the PLA. Chinese leaders understand that unhindered access to the Internet and foreign media will have severe consequences on the nation's political stability and the existing system of governance. Thus the CPC not only wages IW against its adversaries, it also does so against its own population.

China also sees US domination of public information and the Internet as a tool for conducting IW. Some of the biggest hardware and software firms like IBM, Intel, Apple, Dell, Hewlett Packard, Cisco, Microsoft, Oracle etc are all US multinationals. Some of the biggest media companies in the world like Comcast, Walt Disney, Time Warner, Fox Entertainment Group and CBS Corporation are similarly US based. Big Internet companies like Google, Amazon and Facebook are likewise US conglomerates. Together these firms control a huge chunk of the information battlespace be it in terms of generation, transmission, storage, aggregation, presentation, interpretation of information or the underlying infrastructure. The USA does not directly control these ICT conglomerates but the secret documents leaked by former US NSA contractor Edward Snowden, particularly revelations about the PRISM surveillance program are a clear indicator of the extent of collusion existing. The role of western media in justifying the US invasion of Iraq on pretext of non-existent weapons of mass destruction is well known and a clear example of IW conducted by non-state actors. The US also exercises significant control over the Internet both in terms of its architecture, technical governance and regulatory framework.

Internet being a global common and the predominant means of exchanging information, the Communist Party of China (CPC) accords highest priority to controlling not just the content but also the architecture of Internet. To achieve this, China has actively collaborated with other nations to call for a reduction of USA's role in governing the Internet; especially after USA's covert global surveillance program was exposed. China alongwith Russia have also argued for greater autonomy to sovereign nations in determining the Internet architecture and content within their domiciles[13,14,15,16,17].

At the operational (technical) level, China has engineered what is popularly called as the Great Firewall of China to control the flow of information within its territory. Most non-Chinese Internet websites/applications/services (like Youtube, Facebook, Twitter, Instagram, Pinterest, Tumblr, Snapchat, WordPress, Blogspot, Blogger, Flickr, SoundCloud, Netflix, Google+, Google Hangouts, Google Play, Whatsapp, Telegram, Vimeo, Google Drive, Google Docs, Gmail, Google Calendar, Dropbox, ShutterStock, Slideshare, Scribd etc) are blocked thus forcing its citizens to use Chinese equivalents. China also blocks media outlets like The New York Times, Finacial Times, The Wall Street Journal, The Economist Bloomberg, Reuters, The Independent, LeMonde, Google News etc and also certain information repositories like Wikipedia and Wikileaks[18]. The Chinese media on the other hand is state controlled. The Chinese are also investing heavily into media and technology firms across the world so that they have a greater say in the content and presentation of information.

China has also leveraged its position as the leading manufacturer of smartphones, electronics and telecommunication to expand its footprint in the global information infrastructure similar to US. In November 2016 US security firm Kryptowire discovered that a secret software which came preinstalled on low cost Android smartphones was transmitting full contents of text messages, contact lists, call logs, location information and other data to a Chinese server every 72 hours without user consent or knowledge[19]. The Chinese firm that wrote the software, Shanghai Adups Technology Company, says its code runs on more than 700 million phones, cars and other smart devices and that it intentionally designed the software to help a Chinese phone manufacturer monitor user behavior. According to the report, Adups provides software to two of the largest cellphone manufacturers in the world, ZTE and Huawei[20]. While prima-facie the collection of user's private data by Chinese firms is no different from what Google, Facebook and other American ICT conglomerates do, the difference lies in that the PLA is deeply entrenched into the Chinese R&D and manufacturing ecosystem and thus the likelihood of data being exploited for military purposes is an order of magnitude higher.

On the offensive side, China recognizes the unique asymmetric potential of the Internet as an opaque battlespace where covert non-state actors can undermine USA (and other nations) without risking escalation into a kinetic war. Accordingly China (alongwith Russia) has steadfastly objected to the Tallinn process initiated by NATO which tried to lay down global norms for cyber warfare and restrict China's capability[21].

The reason for including this section under the IW capabilities of China is that the extensive regulatory and technical control mechanisms put in place by China give it a definitive defensive edge when it comes to countering offensive IW operations launched by its adversaries. Likewise, China's position as one of the world's leading telecommunications hardware manufacturer give it a launchpad to unleash IW attacks against the critical infrastructure of its adversaries.

IW Force Structure of China

Chinese military philosophy lays great stress on the use (and manipulation) of information for achieving strategic and political objectives, however in the modern battlefield it is technological capabilities and an organized well trained force which matter more. While complete details of PLA's IW headquarters and units are not readily available, a summary of the known capability is presented in this section.

The Armed Forces of China, collectively called the People's Liberation Army (PLA) are controlled by the Central Military Commission (CMC) which in turn is an organ of the Central Committee of the CPC. Until 2015, the next level of hierarchy below the CMC comprised four departments namely the General Staff Department (GSD), the General Political Department (GPD), the General Logistic Department (GLD) and the General Armaments Department (GAD). The Forces themselves were organized into seven Military Regions (MRs), namely Shenyang, Beijing, Lanzhou, Jinan, Nanjing, Guangzhou and Chengdu. The Chinese missile forces were consolidated under the 2nd Artillery Force. The PLA Ground Forces (PLAGF) were controlled directly by the GSD while the PLA Navy (PLAN) and PLA Air Force (PLAAF) had their respective headquarters. The GSD itself was dominated by officers from PLAGF.

In September 2015, China announced that its armed forces would be cut down by 300,000 troops from the original 2.3 million. Three new General Command Headquarters namely the PLA Army (the Ground Forces were earlier directly under the GSD), the PLA Rocket Force (an elevation and renaming of the 2nd Artillery Force), and the PLA Strategic Support Force (a new organization) were established in late 2015. Creation of a Joint Logistics Support Force was announced in 2016. Also, the four military departments (GSD, GPD, GLD and GAD) were disbanded in 2016 and replaced by 15 functional agencies (seven departments/offices, three commissions and five directly affiliated bodies) directly under the CMC[22,23]. These 15 new (or revamped) organizations are as under:

- General Office
- Joint Staff Department
- Political Work Department
- Logistics Support Department
- Equipment Development Department
- Training and Administration Department
- National Defense Mobilization Department
- Discipline Inspection Commission
- Politics and Law Commission
- Science & Technology Commission
- Office for Strategic Planning
- Office for Reform & Organizational Structure
- Office for International Military Cooperation
- Audit Office
- Agency for Offices Administration

Likewise the seven Military Regions were reorganized into five theatre commands, namely the Eastern, Western, Northern, Southern and Central theatre commands. President Xi at the opening session of the 19th CPC National Congress stated that "*mechanization will be basically achieved, with IT application coming a long way and strategic capabilities seeing a big improvement by the year 2020. The modernization of the national defense and armed forces should be basically completed by 2035*"[24].

It is also necessary here to underscore that the PLA is increasingly becoming an expeditionary force similar to the US Armed Forces. Over the past few years, China has not only asserted itself in its neighborhood; it has created (and is in the process of creating) a number of military bases to protect its strategic interests and sea lanes of communication in the Indian Ocean and Pacific theatre. As of now PLA has not fought any wars overseas but it views such expeditionary capability as an increasingly essential element of comprehensive national power and China's ability to challenge US hegemony.

We must also not forget that the PLA is much more of a political organization then its counterparts in other countries and it is essentially an extension of the Communist Party. Achieving jointness and combined arms synergy therefore appears to be the main reason behind the force restructuring; however there are political, economic and corruption related considerations behind it as well. The changes are ongoing and further refinements can be expected as the new organizations mature[25,26,27].

The geographical area of responsibility of the newly created theatre commands still remains centered on the Chinese mainland and adjacent seas (unlike the US European, Pacific and Africa commands), thus implying that they do not have a global orientation. This too may change in due course.

Let us first try to list out the IW force structures of PLA before getting into details of the reorganization that is taking place and its implications.

General Staff Department

The PLA General Staff Department (GSD) was a 1st Level Organization headed by a General rank officer. The GSD in turn comprised of several Departments (2nd Level/Corps Leader). Each department in turn had several Bureaus (3rd

Level/Division Leader) and Bureaus had Offices (4th Level/Regiment Leader)[28]. A brief and generic description of the GSD Departments is as under:

- General Office
- Operations department (Also called the First Department). Post reorganization this has most likely moved under the newly created JSD.
- Military Intelligence department (Also called the Second Department).
- Department of Technical Reconnaissance (Also called the Third Department/3PLA)
- Informatization Department (the erstwhile Communications Department which was also headquarters of the PLA Signal Corps was renamed as Informatization Department in 2011)[29,30]
- Electronic Countermeasures and Radar Department (also called the Fourth Department/4PLA and Department of EW. Some confusion exists whether this or the Communications Department is the Fourth Department).
- Military Affairs Department (similar to Adjutant General's branch)
- Military Training Department
- Mobilization Department (responsible for PLA Reserves and Militia)
- Political Department
- Service Arms Department (it had bureaus corresponding to arms except Infantry and Signals; it also had an Armament Bureau, the Anti-Chemical Warfare Bureau and Special Operations Forces Bureau)
- Independent Bureaus like the Army Aviation Bureau, Management Bureau and Guards Bureau

Each of the erstwhile Military Regions also had Departments that mirrored the GSD. It needs to be emphasized that GSD Departments were not Staff Headquarters and the subordinate bureaus and offices were more akin to units. The departments collaborate with each other in areas of shared interest[31].

PLA Units use five digit tactical numbers (called cover designations) to keep their identity confidential. GSD Units use the cover designation block from 61001 to 61999. The GSD 3rd Department/3rd Bureau/3rd Office is for example Unit 61587. The earlier cover designations have possibly changed after reorganization.

Going into details of each and every Department, their Bureaus and subordinate Work Offices would distract us from our purpose so we shall instead focus only on those Departments of GSD which had an offensive or defensive IW or information related capability:

- **Second Department.** The 2nd Department of GSD (also referred to as the GSD Intelligence Department or PLA Intelligence or 2PLA) was responsible for collection of HUMINT (also referred to as tactical reconnaissance) and also gathered Open Source Intelligence (OSINT). The department also carried out analysis and fusion of gathered intelligence with SIGINT, TECHINT and IMINT to generate finished intelligence products. 2PLA also possibly controlled some of the PLA's unmanned aerial vehicles (UAVs) and Special Forces; as well as PLA's imagery satellites[32].

  The 2nd Department had a number of Bureaus and affiliated institutions like the Institute for International Relations and Institute for International Strategic Studies. A tentative list of the Bureaus under 2nd Department alongwith possible responsibilities is as under[33,34,35,36]:

  - 1st Bureau – Taiwan
  - 2nd Bureau – Provided field intelligence units (called tactical reconnaissance bureaus) to the Military Regions
  - 3rd Bureau – Handled the Military Attaches
  - 4th Bureau – Commonwealth of Independent States (CIS) and East Europe
  - 5th Bureau – USA and Western Europe
  - 6th Bureau – Neighboring Asian countries
  - 7th Bureau – Research and technology development
  - Analysis Bureau – Intelligence fusion
  - Confidential Bureau – Handling of secret documents
  - Aerospace Reconnaissance Bureau – Space based ISR

Analysts often equate 2PLA with the US DIA.

- **Third Department.** The 3rd Department of GSD (also referred to as the Department of Technical Reconnaissance or 3PLA) was responsible for SIGINT (also called technical reconnaissance) and monitoring of foreign telecommunications. Similar to US terminology, the term SIGINT has a broad connotation and encompasses a wide range of CNE, ESM as well as cryptologic functions. Apparently manned by thousands of personnel, including linguists and domain experts, the 3rd Department has been assessed to have around twelve operational bureaus responsible for various linguistic regions and capabilities. Eight of the Bureaus were in and around Beijing, two in Shanghai and one each in Qingdao and Wuhan. Each of the Bureaus had several subordinate Work Offices and Work Stations which were spread out geographically, possibly also including foreign countries. The third department also collaborated with telecommunication service providers like the Shanghai City's Internet monitoring center (dubbed the Shanghai 005 Center), which is managed by China Telecom and numerous academic institutions like the Beijing North Computing Center (BNCC). The 3rd Department also oversees the 56th Research Institute (also known as Jiangnan Computer Technology Research Institute and which operates supercomputers), 57th Research Institute (also known as Southwest Institute of Electronics and Telecommunications Technology and which specializes in satellite communications and signal processing) and the 58th Research Institute (also known as Southwest Automation Research Institute and which specializes in cryptology and information security). An approximate description of the known Bureaus of 3rd Department alongwith speculated roles is as given below:[37,38,39,40,41]

- 1st Bureau (61786 Unit/Beijing) – Cryptologic functions
- 2nd Bureau (61398 Unit/Shanghai) – The 2nd Bureau is one of the largest bureaus of the 3rd Department and possibly responsible for English language targets. Its main base is in Shanghai and it has eight Offices, large number of Work Stations and possibly also a mobile group. The main role of 2nd Bureau appears to be CNE, however it also controls several radio interception, direction finding and technical reconnaissance sites at dispersed locations throughout China. It possibly also intercepts and analyses traffic leaving and entering China through its Offices present near fiber optic cable landing points. Most of the offices of 2nd Bureau operate under the cover designator Unit 61398 which has been dubbed as an

APT. In May 2014, the United States Department of Justice (DOJ) indicted five PLA officers of GSD/3rd Department/2nd Bureau/3rd Office for indulging in cyber espionage activities against US Corporates[42].

- 3rd Bureau (61785 Unit/Beijing) – Defensive (information security) functions including evaluation and certification of cryptologic products. Possibly also intercepts V/UHF radio communications. Offices are spread out across China.
- 4th Bureau (61419 Unit/Qingdao) – Japan and Korea
- 5th Bureau (61565 Unit/Beijing) – Russia, possible SATCOMN capability
- 6th Bureau (61726 Unit/Wuhan) – Taiwan and South Asia
- 7th Bureau (61580 Unit/Beijing) – Possibly some information security functions in conjunction with National Research Center for Information Technology Security which is located adjacent, also some satellite related role
- 8th Bureau (61046 Unit/Beijing) – Europe, Middle East, Africa, Latin America
- 9th Bureau – Possibly SIGINT analysis and database management
- 10th Bureau (61886 Unit/Beijing) – Possibly Central Asia and Russia, also telemetry (missile tracking/nuclear testing)
- 11th Bureau (61672 Unit/Beijing) – Russia
- 12th Bureau (61486 Unit/Shanghai) – CNE, also interception of satellite communication and space based SIGINT. Unit 61486 has been dubbed as yet another APT.

The 3rd Department and its bureaus mostly appear to be static units having close working relation with research institutions. The Department also maintains close liaison with telecommunication service providers which enable the requisite cover and deniability so that SIGINT operations are not verifiably traced back and attributed to the PLA.

The 3rd Department has been accused of widespread espionage of business interests, intrusion into networks, theft of intellectual property and sensitive commercial information by the US Government.

The similarities between role and organization of the GSD 3rd Department and the US NSA/CSS are fairly evident.

- **Fourth Department.** The 4th Department of GSD (also called the Department of Electronic Warfare or 4PLA) was responsible for ELINT and ECM. The Bureaus under 4th Department include the Electronic Countermeasures Bureau, Tactical Unit Management Bureau, the Radar Bureau as well as various ECM and Radar Brigades and Regiments in various regions. The 4th Department and its units have the ability to jam radars, satellites and GPS. The department also included/collaborated with several establishments like the Bureau of Technology and Equipment, Institute of Electronic Engineering and the 54th Institute of China Electronics Technology Group which develops and manufactures telecommunication, electronics and IT equipment. The department is also apparently responsible for maintaining the Optoelectronic Countermeasure Room, threat libraries, generation of E-ORBAT, EW Doctrine and Strategy and guidance on ECCM. China has UAVs capable of collecting ELINT and COMINT (viz. the Austrian Schiebel S-100) and 4PLA has been reported to be employing them in support of its mission. Interestingly Major General Dai Qingmin, a former Director of the Communications Department who later rose to head 4PLA is credited with publishing a seminal work on Integrated Network Electronic Warfare (INEW). The INEW doctrine considers EW and CNO as the predominant forms of IW and also dwells into aspects of integration with kinetic means, particularly strategic level attacks against space based targets. Clearly the EW capabilities appear to be consolidated under the 4th Department[43,44,45,46]. Some sources suggest that the 4th Department was also responsible for CNA[47].
- **Informatization (Communications) Department.** The GSD Communications Department, which was also the headquarters of the PLA Signal Corps, was a national level organization also sometimes referred to as 5PLA by western scholars. The department was responsible for deploying, manning and operating the PLAs C4I infrastructure, the administrative communication systems and also operated the PLAs national command and control centre in Beijing. It was also responsible for automation and information security

functions. The department was deeply entrenched in China's commercial telecommunication sector and played a leading role in modernizing it. The department also controlled many R&D institutes and training academies. The possible Bureaus/units under the Informatization Department are as under[48]:

- Strategic Communication Bureau
- Tactical Communications Bureau
- Communications Security Bureau
- Communications Automation Bureau
- Organic Signal Corps Units (possibly for manning the strategic assets)

The renaming of Communications Department to Informatization Department in 2011 indicates that China's new doctrine of fighting wars under conditions of informatization was as much about enabling network-centricity and integration of C4ISR capabilities as it was about EW and cyber warfare. More recently the informatization department has also been reported to be involved in tasks related to sensor data fusion and generation of common operational picture. The department is also reported to be flying UAVs for reconnaissance[49,50,51].

PLA Ground Forces (PLAGF)

The PLAGF have been substantially downsized as part of the modernization efforts. There also appear to be efforts being made to organize them into smaller Brigade level combat forces. While other elements have been downsized; it is reported that reconnaissance, surface-to-air missile, Special Forces, helicopter, and EW elements have all been expanded[52].

In terms of specific units, each of the seven MRs had SIGINT Technical Reconnaissance Bureaus (TRBs) similar to the ones under the GSD. As already brought out, each TRB can be expected to have multiple specialized offices (regiments) deployed across the area of responsibility. The number of TRBs under the erstwhile MRs is as under[53]:

- *Beijing, Guangzhou, Jinan and Shenyan MRs.* One TRB each
- *Chengdu, Lanzhou and Nanjing MRs.* Two TRBs each

If we go into the organization of the erstwhile MRs we observe that they comprised of two or three Group Armies and one Group Army within the MR had an EW Regiment under command[54]. This essentially translates into one EW Regiment per MR, possibly directly under command. The number of units (and their equipment profile) may have improved as part of the ongoing modernization and expansion of the EW elements.

PLA Air Force (PLAAF)

The PLAAF has three TRBs possibly supporting airborne SIGINT collection in addition to monitoring and direction finding of a wide range of communications[55]. PLAAF's modernization efforts in recent years are focused not only on development of advanced fighter jets with stealth capability but also on the development and deployment of support aircraft serving as force multipliers to enhance the effectiveness of its combat aircraft. These support aircraft include tankers, AWACS, EW, intelligence collection and search & rescue aircraft[56]. Air Force specific SIGINT collection is managed by the PLAAF Sixth Research Institute in Beijing[57].

In terms of EW aircrafts, the PLAAF reportedly has several Shaanxi Y-8 variants which can conduct ECM (Shaanxi Y-8C) and SIGINT tasks[58].

The newly inducted Shenyang J-16 fighter aircrafts has an EW variant (J-16D) for SEAD role similar to the US EA-18G Growler[59].

PLA Navy (PLAN)

Besides the EW capability inherent in the naval platforms themselves, the PLAN reportedly has two dedicated TRBs[60]. Since the PLA comprises of three fleets, namely the North Sea Fleet, the East Sea Fleet, and the South Sea Fleet whereas the number of TRBs is only two, we can assume some sort of resource sharing/splitting arrangements to be in place.

PLAN also operates the newly inducted H-6G EW aircraft[61].

In addition to above, the newly inducted Shenyang J-15 fighter aircrafts are also capable of being configured for EW role[62].

PLA Rocket Force (PLARF/2nd Artillery)

The PLA RF has one TRB[63].

Marine Corps

The Chinese Marine Corps (reorganized in year 2006) is reported to have an integral Communications, Information and EW Group[64]. It was also speculated that two recent collisions between US Navy warships and commercial ships were engineered by Chinese EW units subverting or deceiving the radar and navigation systems of the ships[65]. How far this is true is not known but it is technically feasible.

General Political Department (GPD)

While SIGINT, Cyber Warfare and EW capabilities seem to be vested in the GSD and the field units of PLAGF, PLAN and PLAAF; the Psy Warfare functions were apparently controlled by the GPD. Without going into details of the many departments and bureaus within the GPD, it is broadly responsible for propaganda, ideological, organizational and disciplinary work of the PLA as also liaising with foreigners. The reason for placing Psy Warfare under the GPD is possibly because it was also responsible for propaganda and the two are intimately related. Within GPD it is the Liaison Department which handled propaganda against enemy forces and relations with Taiwan. The Liaison Department comprised of at least four Bureaus whose functions were as under:[66]

- **Liaison Bureau.** Operations pertaining to Taiwan
- **Investigation and Research Bureau.** Security analysis and friendly contact
- **External Propaganda Bureau.** Disintegration operations to include Psy and Legal Warfare
- **Border Defense Bureau.** Border negotiations and agreements

One specific unit called the 311 Base (possibly also referred to as Unit 61716/ 31st Base/310 Unit) established under the Liaison Department in 2005 is reportedly mandated to operationalize the Three Warfares strategy. 311 Base is headquartered at Fuzhou in Fujian province[67].

The 311 Base possibly commands six regiments and its main focus has been on Taiwan. The Commander of Base 311 (Major General rank) is possibly also dual-hatted as the CEO of the China Huayi Broadcasting Corporation (CHBC)[68].

The 311 Base also has affiliation with a number of organizations which promote Chinese culture in foreign countries. These include the China Association for Promotion of Chinese Culture (CAPCC); China Association for Friendly International Contacts (CAIFC); China-US Exchange Foundation (CUSEF), The Centre for Peace and Development Studies (CPDS), External Propaganda Bureau (EPB), and China Energy Fund Committee (CEFC)[69].

In addition to the 311 Base, the Chinese military has possibly established psy warfare units in all erstwhile MRs, set up psy warfare specialties in military academies, and also established a psy warfare institute. Psy warfare officers have been assigned to serve in the armed forces of land, sea and air[70].

Ministry of State Security (MSS)

The Ministry of State Security (MSS) is the external intelligence and security agency of China. It is responsible for counter-intelligence, foreign intelligence and political security. Unlike the US CIA, the MSS also has a law enforcement charter and it conducts both foreign and domestic espionage activities. The dominant role played by the PLA intelligence apparatus suggests that the MSS is mostly involved in non-military intelligence tasks, particularly economic espionage. HUMINT remains a key area of focus, but the MSS has also forayed into cyber espionage and has been dubbed an APT by US security analysts[71].

Space Capabilities of China

Space and Information are the two new battlespaces in addition to the conventional ones of Land, Sea and Air. While information modulates each one of the battlespaces, it is in case of space that its impact is most pronounced and well defined. The reason for this is that space is mostly an unmanned arena and entities deployed in this battlespace must per force use wireless links and electronic computational devices (hardware and software) for command, control, telemetry and relaying back data. This makes space based assets immensely vulnerable to EW and Cyber Warfare. In fact the non-destructive, reusable, simple and inexpensive nature of these IW techniques combined with the advantage of deniability makes them the preferred method of attack vis-à-vis kinetic means.

What also merits attention is that space based assets lie at the heart of not just long haul communications but also ISR, position, navigation, and timing

(PNT) and ballistic missile defense capabilities which are vital in the modern battlefield. Any substantial degradation of these capabilities has the potential of changing the entire operational dynamics and severely downgrade the adversary's warfighting potential. Expeditionary forces like the US and offensive formations which fight deep within enemy territory are the most vulnerable.

With this preamble we shall take a brief look at the space capabilities of China. The reason we are doing so (and did not do so in case of US) is because the SSF aims to bring together both the space and IW capabilities of the PLA under a single umbrella. We must therefore broadly understand what the space related capabilities are before we examine the SSF itself.

Vision

USA presently has the most advanced space program in the world. More importantly, it has fully integrated its space based communication, surveillance and navigation capability with its military, thus giving it the ability to rapidly project power anywhere in the world. The US also uses space for ballistic missile defense and intelligence gathering. In order to match this formidable capability, China has again adopted a two pronged approach. Firstly, China is trying to catch up with US in space crafts and satellites at the earliest. Secondly China is investing in anti-satellite (ASAT) capability to act as a deterrent and also provide it with an offensive option.

In adherence to this approach China today possesses the most rapidly maturing space program in the world with particular emphasis on satellite communication (SATCOM), ISR, satellite navigation (SATNAV), PNT, meteorology; as well as manned, unmanned and interplanetary space exploration. The PLA is one of the major beneficiaries of China's burgeoning space program and allows it to not only monitor objects across the globe as well as in space but also expand its command and control umbrella. PLA strategists regard the ability to use space based systems and deny adversaries access to the same as central to enabling modern, "informationized" warfare.

Satellites and Launch Vehicles

China now has approximately 193 operational satellites in orbit, with approximately 95 of these owned and operated by military or defense industry organizations. China's current system of C4ISR satellites enables its military to detect and monitor air and naval activity in the Indian Ocean and Pacific region with sufficient accuracy and timeliness to assess military force posture and direct other assets for more precise tracking and targeting. China's BeiDou system is expected to achieve global coverage by 2022 and will provide PNT functions, essential to the performance of virtually every modern Chinese weapons system, independent of US run GPS.

In addition to an array of space launch vehicles and the Tiangong-1 prototype space station, China is also developing advanced dual purpose spacecrafts such as the Shenlong, which is similar to the US Air Force's mysterious X-37B space plane. Shenlong reportedly embarked on its first test flight in 2011 and is designed to take off like a rocket and land like a shuttle. It boasts of high speed, maneuverability and radar-evading stealth features. It will be capable of long-range flight and can serve as a space weapons launch platform, as well as for surveillance, intelligence and early warning missions[72].

The Chinese aerospace industry is concurrently developing two reusable launch vehicle (RLV) systems. The first one is a crewed space shuttle launched vertically atop the Changzheng 5 (CZ-5) rocket. The second is a smaller unmanned suborbital space plane that can launch a second-stage rocket to deliver payload to orbit. The unmanned system is expected to enter service before 2020 to supplement the country's existing range of conventional non-reusable launch vehicles. The crewed space shuttle may replace the existing Shenzhou human capsule for crew transportation to the future space station by 2030[73].

China Aerospace Science and Industry Corporation (CASIC) is the main contractor for China's space program and also its largest missile designer and manufacturer. The organization plans and oversees the development, production, and testing of China's direct ascent anti-satellite assets and operationally responsive launch capability, including the associated road-mobile launchers and small satellites. CASIC employed more than 135,000 workers in 2013[74].

Denial of Space (Counter-Space)

China continues to develop a variety of capabilities designed to limit or prevent the use of space based assets by adversaries during a crisis or conflict, including the development of directed-energy weapons, satellite jammers and capabilities to interdict satellites with ground-based missiles.

In 2007, China carried out a successful test of its first ASAT weapon, destroying an aging Chinese weather satellite Fengyun-1C (FY-1C) in Low Earth Orbit at an altitude of 864 Kms. The test demonstrated China's ability to strike satellites in low Earth orbit, where majority of the spy satellites reside. The missile was fired from a mobile transporter-erector-launcher from a launch site at the Xichang space facility in Sichuan province. This demonstration of ASAT capabilities was remarkable as the attack was executed on a spacecraft that was flying at 7.42 km per second, similar to an intercontinental ballistic missile re-entering the atmosphere. China has tested two direct-ascent ASAT missiles – the SC-19 and the larger Dong Neng-2 (DN-2). Direct-ascent ASAT missiles are designed to disable or destroy a satellite or spacecraft. China conducted additional SC-19 tests in 2010, 2013, and 2014. In each test, the SC-19 intercepted a mock warhead launched by a ballistic missile[75,76,77].

Although China has called these tests "land-based missile interception tests," available evidence suggests they were indeed ASAT tests. In May 2013, China fired its new DN-2 rocket into nearly Geosynchronous Earth Orbit ascending to around 30,000 kilometres (18,600 miles) above earth[78]. This indicates that China is developing the capability to target higher orbits which contain PNT satellites. Beijing claims the launch was part of a high-altitude scientific experiment; however, the nature of the test indicates that China is developing an ASAT capability to target satellites in medium earth orbit, highly elliptical earth orbit, as well as the geosynchronous earth orbit. On October 30, 2015 China conducted flight test of a new ASAT missile DN-3 from the Korla Missile Test Complex in western China[79].

China has also embarked on a program to develop co-orbital satellite interceptors, launched from earth into a temporary parking orbit from which they then maneuver to their target. Robotic arms on such interceptors can be used for civilian missions such as satellite repair, space station construction, and orbital debris removal; but these can also target a satellite to perform various ASAT missions[80]. Options include monitoring, disabling/destroying the target satellite using jamming, directed energy weapons, robotic arms, collision, explosive charges and/or kinetic projectiles. Co-orbital ASAT systems provide several advantages over current direct-ascent missiles, including their ability to target satellites in every orbital regime, generate less debris, conduct attacks without geographic limitations, and limit escalation, as many co-orbital attack options are reversible and offer plausible deniability. Such a capability would give the Chinese military three significant benefits. Firstly, it would allow attacks on spacecraft whose orbital tracks might not normally traverse the Chinese mainland; secondly it would provide a covert 'sleeper' space attack option that could unfold over a period ranging from hours to days to months, unlike direct-ascent systems whose operations are overt and conclude in a matter of minutes. Lastly it would provide insurance for ASAT attack options in the event direct-ascent systems were destroyed early in a conflict[81,82].

During a manned space mission in September 2008, China's Shenzhou 7 spacecraft deployed the BX-1, a miniature imaging satellite, which then positioned itself into an orbit around the spacecraft. At one point the BX-1 passed within 45 kilometers of the International Space Station, apparently without prior notification, suggesting it may have been simulating a co-orbital ASAT attack. In June 2010, China launched the SJ-12 satellite. Over the next two months, the satellite conducted a series of maneuvers and came within proximity of the SJ-6F, an older Chinese satellite that was placed into orbit in 2008. The activities of the SJ-12 may have been designed to test a co-orbital ASAT capability, such as on-orbit jamming. Moreover, during its maneuvers, the SJ-12 apparently bumped the SJ-6F, causing it to drift slightly from its orbit. This activity suggests China also could have used the test to demonstrate the ability to move a target satellite out of its intended position by hitting it or attaching to it. In July 2013, China launched a rocket carrying the CX-3, SY-7, and SJ-15 satellites, one of which was equipped with a robotic arm for grabbing or capturing items in space. Once all three were in orbit, the satellite with the robotic arm grappled one of the other satellites, which was acting as a target satellite. The satellite with the robotic arm then changed orbits and came within proximity of a separate satellite, the SJ-7, an older Chinese satellite that was sent into orbit in 2005[83].

China's Small Satellite Research Institute of the Chinese Academy of Space Technology (CAST) is also developing a "parasitic satellite" which when deployed gets attached to the enemy's satellite; and during conflict, commands are sent to it which will than interfere with or destroy the host satellite. Since the "parasitic satellites" reside with their host and are only activated during a conflict, their volumes and masses must be very small to conceal their existence and this requires significant miniaturization and innovation. Ground testing has shown that "parasitic satellites" are very effective and efficient in their operations. When a "parasitic satellite" is properly deployed, it would disable or destroy the host satellite system in less than a minute. The cost of building a "parasitic satellite" is 0.1 to 1 percent of a typical satellite, thus its deployment is highly cost effective. When China's new reusable two-stage launchers become available in the near future, the deployment cost of "parasitic satellites" will be lowered further[84].

Summary

To summarize, China's space program with military connotations has improved both quantitatively and qualitatively, launching more satellites with more advanced capabilities each succeeding year. In particular, they have repeatedly tested ASAT systems, recently demonstrating capabilities that would threaten even high altitude US early warning satellites. China is also testing increasingly complex co-orbital proximity capabilities. Although it may not develop or operationally deploy all of these co-orbital technologies for counter space missions, China is setting a strong foundation for future co-orbital ASAT systems that could include jammers, robotic arms, kinetic kill vehicles, and lasers. The Chinese vision of space warfare involves not just denying space to its adversary but also using space for affirmative ends such as the intercept of ballistic and cruise missiles through space-based combat platforms, strikes by space systems on terrestrial targets and attacks by land, air, sea, aerospace and space vehicles on an adversary's space platforms and space-based command and control assets.

Having seen both the IW and Space capability, let us now get to the PLA Strategic Support Force.

The Strategic Support Force

One of the most talked about and interesting element of the Chinese military reforms has been creation of the PLA Strategic Support Force (PLA SSF) and its subsequent elevation to be at par with the Ground Forces, Navy, Air Force and Rocket Force.

The SSF is a nascent organization which is still evolving and details about its operational capabilities and employment are not entirely clear. However, knowledge of the elements subordinated to SSF helps us to analyze the underlying IW philosophy and how they foresee integration of IW capabilities with the other four battle spaces of land, sea, air and space as also with kinetic means.

The stated purpose of the SSF is to unify the C4ISR, space, cyber, and EW capabilities in support of operations by joint theatre commands. The 2013 edition of the Science of Strategy alludes to a more offensive orientation stating that the PLA should integrate space, cyber, and electronic warfare operations to "paralyze enemy operational systems". The SSF is known to comprise of two main departments (verticals) namely the Space Systems Department and the Network Systems Department. Broadly speaking this signifies a consolidation of the space and IW capabilities.

Official Chinese press releases and articles of 2016 indicate that the mission of the SSF is to ensure that the PLA can maintain superiority in military operations taking place in space (i.e. ISR, PNT, counter-space) and involving the electromagnetic field (i.e. EW) and the Internet (i.e. SIGINT/Cyber Warfare). The force's responsibilities include target reconnaissance and tracking, global positioning operations, space assets management and defending against hostile activities in cyberspace and the electromagnetic field. Researchers of the SSF are also supposed to be focusing on cutting-edge technologies such as big data applications, cloud computing, 3-D printing and nano-materials[85].

In 2017 China announced that it had adjusted or established 84 Corps level units as part of its ongoing reorganization[86]. While exact details of the reshuffle were not disclosed, it is obvious that several of these would pertain to the newly created SSF.

Over the past two years, think tanks and analysts have tried to assess the role, organization and tasking of SSF and a hazy picture is beginning to emerge. Almost all the departments and units of PLA that we have covered above were candidates for moving to the SSF, however that does not seem to have happened.

Composition of the SSF

In an update published in the 2016 China Brief of The Jamestown Foundation, the author John Costello opines that the Space Systems Department (SSD) of the SSF consolidates the space launch, support, telemetry, tracking, and control (TT&C) and intelligence, surveillance, and reconnaissance (ISR). 3PLA and 4PLA (possibly less some elements) have moved to the Network Systems Department (NSD). The Informatization Department on the other hand has not moved to the SSF and instead has been reorganized as the Information Communications Bureau under the newly created JSD. However both the

Satellite Main Station from the Informatization Department and the Aerospace Reconnaissance Bureau (ARB) of 2PLA have been incorporated into the SSF[87].

In a subsequent testimony given to the US-China Economic and Security Review Commission in February 2018, the same analyst opines that the 3rd Department (3PLA) has moved to the SSF Network Systems Department in its entirety. As regards the Space Systems Department he states that "*It is currently unclear what responsibilities, if any, the SSF's space corps has for anti-satellite missile (ASAT) research, development, testing, and operations, nor is it known whether the SSF has a role in the related discipline of ballistic missile defense (BMD). Both missions could presumably fall under the categories of space attack and defense, respectively, which would place them under the Strategic Support Force's remit. Alternatively, these missions may be assigned to the PLA Rocket Force, which already has a role in missile operations, or the PLA Air Force, which have already demonstrated a limited capability in both anti-satellite missiles and ballistic missile defense*"[88].

The same view is shared in a 2018 report by the US Center for Strategic and International Studies (CSIS) as per which SSF does not have full authority over the nation's arsenal of direct-ascent ASAT weapons, however it does appear to have control over some of the other types of counter-space capabilities[89].

In an article of January 2017, news and analysis site sinodefence.com stated that SSF has created two space related organizations under it, namely the Space System Department (SSD) and the Military Space Force (MSF). The article contends that SSD is possibly a renaming of China Satellite Launch and Tracking Control General (CLTC) which was established in 1986 to function as the command and control centre for space-related operations of the PLA and was headed by the Chief of Staff of erstwhile GAD. The article also notes that China's manned spaceflight program has not moved to SSF and is under the newly constituted Equipment Development Department[90].

Both Costello (Jamestown Foundation) and the Sinodefence website give out almost identical ORBAT of the MSF as comprising of satellite launch and satellite tracking/telemetry/control centers as under:

- Jiuquan Satellite Launch Centre (Base 20)
- Taiyuan Satellite Launch Centre (Base 25)
- Xichang Satellite Control Centre (Base 27)

- Wenchang Aerospace Launch Site
- China Satellite Maritime Tracking Control Department (Base 23)
- Xi'an Satellite Control Centre (Base 26)
- Beijing Aerospace Control Centre (BACC)
- China Aerodynamics Research and Development Centre (Base 29)
- Beijing Institute of Tracking and Communication Technology
- Space Research and Development Centre
- Engineering Design Research Institute

As regards the IW component the website states that two organizations namely the Network System Department (NSD) and a branch known as the Network and Information Operations Force (NIOF) have been created under the SSF. Out of these NSD is most likely the erstwhile 3PLA. The website states that 4PLA has also possibly been taken over by SSF. As regards the Informatization (Communications) Department, the website believes that the department has been abolished and replaced by the Information and Communications Bureau under the JSD. However, unlike the previous department, the Information and Communications Bureau is only responsible for maintenance of the PLA's national level C4I system, and does not have the responsibility for overseeing the PLA's communications and automation. As a result, some of these elements may have been transferred to the PLASSF[91].

In an article titled "*PLA Strategic Support Force: The 'Information Umbrella' for China's Military*" in April 2017, analyst Elsa B Kania shares similar views and says that the SSF appears to have taken on a central role in the management of the PLA's BeiDou navigation satellites[92].

A 2016 article on Chinese website Sina.com predicts that the Shenlong orbiter and the Cyber forces of the PLAAF will also be transferred to the SSF. It is not clear if this has actually happened[93].

In a report published in March 2017, scholars from the Institute of National Strategic Studies (INSS), US National Defense University opined that the GSD Operations Department (1PLA) and Intelligence Analysis (2PLA) were possibly assigned to the newly created Joint Staff Department (JSD) whereas 3PLA and 4PLA were possibly merged and moved to the SSF[94].

A 2017 assessment by Jane's echoes the above view opining that 3 and 4 PLA have been incorporated into the SSF whereas 2PLA has possibly moved to the JSD[95].

In a research article published in The Cyber Defense Review edition of Spring 2018, analysts Elsa B Kania and John Costello referred to earlier opine that 3PLA alongwith its TRBs and the two EW Brigades of 4PLA have moved to SSF. The 56th, 57th and 58th Research Institutes affiliated to 3PLA and the 54th research Institute affiliated to 4PLA have also moved to SSF. The authors opine that while the EW Brigades of 4PLA have been subordinated to SSF, the 4th Department itself has moved under the JSD and not the SSF. It is not clear if the TRBs of the erstwhile MRs, PLAN and PLAAF have moved to the SSF. The EW capability thus appears to have been divided amongst these various elements (viz JSD, SSF and Theatre Commands). Lastly, the authors aver that the Informatization Department's Information Assurance Base and its subordinate Network Security and Defense Center has not become part of the SSF and instead is under the newly created JSD's Information and Communications Bureau[96].

The Golbal Taiwan Institute likewise mentions in an article that both the SSF and JSD could field UAVs with EW payloads[97].

A large number of UAVs (including several with EW/ELINT capability) are held by the PLAAF and these do not appear to have been transferred to the SSF[98].

As regards the Psychological Warfare domain (also referred to as Political Warfare in Chinese context), most analysts agree that the 311 Base of the erstwhile GPD/Liaison Department has been shifted under the SSF and integrated with the cyber warfare forces[99,100,101,102,103].

The likely PLA SSF organization as discerned from the above secondary sources is summarized in the table given below:

Sl. No.	*Organization*	*Likely Destination Post Reorganization*	*Elements Subordinated to SSF (if any)*	*Remarks*
			GSD	
1	2PLA	JSD	Aerial Reconnaissance Bureau	ISR capability shed to SSF but HUMINT capability may not be with SSF

Sl. No.	*Organization*	*Likely Destination Post Reorganization*	*Elements Subordinated to SSF (if any)*	*Remarks*
2	3PLA	SSF	All	SIGINT and cyber warfare capability (less that held with MRs, PLAN, PLAAF, PLARF) shed to SSF
3	4PLA	JSD	Radar ECM Brigades and possibly some other units	SSF has some tactical and strategic EW capability (possibly non-kinetic ASAT)
4	Informatization Department	JSD	Few elements may have been shed to SSF	Communications (defensive IW) mostly remains outside the purview of SSF
		Army, Navy, Air Force and Marine Corps		
5	TRBs of MRs, PLAN, PLAAF, PLARF	Theatre Commands	-	Each service retains SIGINT capability and it has not been centralized
6	EW and Communication Regiments of MRs and Marine Corps	Theatre Commands	-	Tactical EW and Defensive IW (COMSEC) functions are not consolidated under SSF
7	EW Aircraft of PLAAF, PLAN	Remain with PLAAF and PLAN	-	SSF may not have SEAD and airborne EW capability
		Other Departments and Organizations		
8	GPD	Core functions of the GPD have been transferred to the new CMC Political Work Department	311 Base	SSF has some Psychological Warfare capability (particularly Taiwan centric) but not complete and while it is integrated with Cyber Warfare, it may not be integrated with HUMINT (2PLA)
9	GAD	Core functions of the GAD have transferred to the new CMC Equipment Development Department	Satellite launch, tracking, telemetry and control centers have been transferred to SSF. SATNAV and space orbiter program have also possibly been transferred to SSF	ASAT and BMD capabilities of PLAAF and PLARF have most likely not been transferred to SSF.
10	GLD	Core functions of the GLD have been transferred to the new CMC Logistics Sustainment Department	-	-
11	MSS	No change	-	MSS SIGINT/Cyber Warfare capabilities have not been merged with SSF

Many of the above observations are supported by websites which transliterate Chinese sources viz the GHQ Vocabulary List[104].

Inferences from the IW Force Structure of China

A number of conclusions can be drawn from the manner in which the IW capabilities of China have been reorganized under the SSF. These are listed below:

- **Defensive IW Operations.** The defensive IW capabilities have not been fully consolidated.
- **Electronic Warfare.** Only limited EW capability has been placed under the SSF. Bulk of the EW capability remains integral to the unified commands and respective branches. Capability subordinated maybe non-kinetic ASAT.
- **Psychological Warfare.** Limited psy warfare capability has been placed under the SSF. Bulk is still integral to unified theatre commands and other departments.
- **Cyber Warfare.** SIGINT and Cyber Warfare capability has been consolidated under the SSF.
- **Intelligence Capabilities.** Limited ISR capabilities have been grouped with the SSF. Bulk of it is still integral to unified theatre commands and other departments.

While the organization is still evolving and details are lacking, the broad characteristics of SSF are fairly evident and summarized below:

- **Offensive Orientation in the Information Domain.** The design and intent of SSF in the information domain is offensive. Incorporating SIGINT units of 3PLA gives SSF a substantial capability to launch Cyber Warfare attacks against enemy's critical infrastructure including the information grid. Incorporation of EW units of 4PLA likewise gives an offensive orientation to the SSF. The defensive IW elements (viz Informatization department and communication units) on the other hand have been left out of the SSF ORBAT except for SATCOM. Possibly the only communication elements subordinated to SSF are the ones required for its own command and control.

- **Defensive Orientation in the Space Domain.** Operation of space based SATCOMN, ISR and PNT assets has been consolidated under the SSF. However, the SSF does not appear to command any kinetic ASAT systems nor is it responsible for BMD. The role of SSF thus appears to be defending these capabilities against attacks and to act as a facilitator for integration, enhancement and standardized of ISR and PNT capabilities across PLA's other service branches.
- **Non-Kinetic Orientation.** SSF does not appear to have any hard-kill capability including ASAT. Thus the SSF does not directly integrate IW with kinetic means, it only facilitates the same.
- **Limited EW Capability Possibly Restricted to ASAT.** The EW regiments of MRs are not under the SSF. EW aircrafts of PLAAF and PLAN are also not part of SSF. The SSF also does not command any substantial portion of EW capable UAVs a lot of which are held with PLAAF. The SSF therefore does not appear capable of conducting any major offensive EW operations like facilitating SEAD. The SSF does inherit some EW units of 4PLA and it would be logical to presume that these could be the ground based non-kinetic ASAT jammers.
- **Integration of Cyber and Psy Warfare.** Incorporation of 311 Base into SSF indicates a certain degree of integration between Psy Ops and Cyber Warfare. The entire Psy Warfare resources however do not appear to have been consolidated under the SSF.
- **Separation of Strategic and Tactical Military Intelligence.** Consolidation of SIGINT and space based ISR capability indicates a consolidation of the strategic military intelligence capability in the technical domains.

It is clear that the SSF is not a merger of the entire IW capability or the entire space capability of PLA as some would like to believe. Most of the tactical/ integral resources appear to have been left out and the SSF seems to be focused mainly on strategic aspects. It is a bigger experiment as compared to the US Cyber Command and has both defensive and offensive connotations. One of its main roles appears to be denial of space and the global information grid through non-kinetic means but it also appears designed to speed up and consolidate the portion of PLAs informatization process which relies on space based assets. This is fairly logical considering that this is the Achilles heels of the US expeditionary forces which may threaten China. If the PLA can ensure survivability of its own space based networking and ISR capabilities while denying the same to the US, the mission of the SSF is accomplished.

It would be premature to comment on whether the SSF is optimally organized because the organization is still evolving. There exist some areas of challenge and these are listed below:

- If the SSF does not have any hard kill capability, then a coordination mechanism between kinetic and non-kinetic means is required when it comes to attacking strategic targets (viz critical infrastructure and space based assets).
- Creation of the SSF is governed by both operational and political considerations. While this may be an optimum balance for the CPC, it may not be the best operationally.
- Staff hierarchies inherent in apex level organizations invariably impose a penalty when it comes to allocation of resources in support of other branches. This is perhaps one of the reasons why the SSF is mostly entrusted with a strategic role.
- There appears to be a fragmentation of the intelligence gathering capability whereby SIGINT and TECHINT are with SSF while the HUMINT resources of 2PLA have gone to the JSD. This is undesirable.
- Psy Warfare couples with both Cyber Warfare as well as HUMINT. Media Warfare and Legal Warfare likewise have multiple dependencies. Consolidating these is a challenge.
- The informatization department having moved under the JSD implies that the SSF Network Systems Department does not have a defensive IW role and these functions will possibly continue to be performed by various other agencies.

Essentially the SSF is a much larger construct as compared to the US Cyber Command which is obvious because it now enjoys the same status as the PLAGF, PLAN, PLAAF and PLARF. While the Space Systems Department (SSD) of the SSF consolidates the space based informatization resources; the Network Systems Department (NSD) comprising of 3PLA and 4PLA elements has an offensive orientation. As of now it appears to be an organization tailor made to counter its main adversary – the US; but it is also a unique experiment in possibly brining the two new battlespaces under a single unified command in the future.

Inferences

The interplay between attack and defense in the information battlespace becomes absolutely clear from the role of the US Cyber Command and PLA SSF. The US forces are expeditionary by design and fight far away from their home bases. The US Cyber Command therefore consolidates their entire defensive might to ensure that their space based assets and global information grid survives against enemy attacks. The PLA SSF on the other hand brings together the corresponding offensive capabilities to attack and bring down the US's space based assets and their information grid. In the long run it may also emerge as the branch which looks after the highly interdependent new battlespaces of information and space just like the Army, Navy and Air Force look after the land, sea and air battlespaces.

It is also obvious that while doctrines lay down a broad concept, the design of forces which carry out these functions is essentially based on their offensive and/or defensive role and the synergy which can be achieved amongst the underlying capabilities and weapon systems. This is again similar to the land battlespace where attack operations require a different type of force which is mobile and has larger number of mechanized elements. The types of elements grouped are such that they match each other in terms of mobility and reinforce each other's capabilities. Defensive operations on the other hand require a different grouping which is capable of holding and defending a piece of ground. The roles of offensive and defensive formations are not interchangeable. Each battlespace therefore requires its own unique and synergistic force structures but they must ultimately be integrated to produce optimum effect.

The defensive US Cyber command and the offensive Chinese SSF are just two illustrative examples of how IW capabilities maybe grouped. We may have many similar permutations and combinations to arrive at different force structures that are complementary and can conduct a broad class of operations in pursuit of these missions. In order to do so we need to lay down which combination of capabilities work and which ones don't. These conclusions are covered in the next chapter.

ENDNOTES

1. The State Council Information Office of the People's Republic of China White Paper on "*China's Military Strategy*" published in May 2015, "Section IV: Building and Development of China's Armed Forces", English translation available at Xinhuanet website, URL: http://www.xinhuanet.com/english/china/2015-05/26/c_134271001_4.htm accessed on 01 May 2018
2. William Carter, "*Chinese Advances in Emerging Technologies and their Implications for US National Security*", (Statement before the US House Armed Services Committee, Subcommittee on Emerging Threats and Capabilities, 2018), p.6
3. Perlroth, Nicole, "*2nd China Army Unit Implicated in Online Spying*", (New York Times, 09 June 2014)
4. FireEye website, "*Advanced Persistent Threat Groups : Who's who of cyber threat actors*", Web URL: https://www.fireeye.com/current-threats/apt-groups.html accessed on 01 May 2018
5. Wikipedia article, "*GhostNet*", Web URL: https://en.wikipedia.org/wiki/Ghostnet accessed on 01 May 2018
6. Alperovitch, Dmitri, "*Revealed: Operation Shady RAT*", McAfee Labs white paper, 2011
7. Hearing Before the Subcommittee on Emerging Threats, Cybersecurity, and Science and Technology of the Committee on Homeland Security, "*Securing the Modern Electric Grid from Physical and Cyber Attacks*", (US Government Printing Office, 2009), p.2.
8. Indian National Critical Information Infrastructure Protection Centre (NCIIPC) Newsletter, January 2017
9. People's Daily Online website, "*PLA Navy deploys 'new type of electronic warfare aircraft*", (Published on 22 January 2018), Web URL: http://en.people.cn/n3/2018/0122/c90000-9417878.html accessed on 01 May 2018
10. Xinhuanet website, "*China displays electronic warfare equipment at Army Day parade*", (Published on 30 July 2017), Web URL: http://www.xinhuanet.com/english/2017-07/30/c_136485220.htm accessed on 01 May 2018
11. Ministry of National Defense of the People's Republic of China White Papers of 2010 and 2014 available at Web URL: http://eng.mod.gov.cn/publications/node_48467.htm accessed on 01 May 2018
12. Wang Xin, "*The Ministry of National Defense responded exclusively to China's "Network Blue Army" and called it a hacker force*", (Published on Huanqiu.com website on 26 May 2011), Web URL : http://china.huanqiu.com/roll/2011-05/1719423.html (as translated by Google Translate) accessed on 01 May 2018
13. Scott Shackelford and Frank Alexander, "*China's cyber sovereignty: paper tiger or rising dragon?*", (Published on Asia and the Pacific Policy Society website on 12 January 2018), Web URL: https://www.policyforum.net/chinas-cyber-sovereignty/?mkt_tok=eyJpIjoiTTJZMU9ESTBPR1F5WVRjeiIsInQiOiJROFpQdnpvS1NVS EJvV1RFcnRuRDdBdkFcL2 tu Z3 BW a2p4aURaR kVxdE0wN3pEamxkY1wvUVwvMmZIVTVvQnFae Dh6Tmp LUm8wbXd QUDluVEFJXC9tYm QyeHVrdDl3ckxHWUVwU1dNNTkwc2 pRaDNG dXdDa0d1Vk 85 SW5URUVkRmRMSXcifQ%3D%3D accessed on 01 May 2018
14. Marika Van Laan, "*ICANN, Russia, China, and Internet Reform: What You Need to Know*" (Published on Ramen IR website on 23 October 2016), Web URL: https://ramenir.com/2016/10/23/icann-russia-china-and-internet-reform-what-you-need-to-know/ accessed on 01 May 2018
15. Alexander Gabuev, "*How China and Russia see the Internet*", (Published on the World Economic

Forum website on 16 December 2015), Web URL: https://www.weforum.org/agenda/2015/12/how-china-and-russia-see-the-internet/accessed on 01 May 2018

16. Yuxi Wei, "*China-Russia Cybersecurity Cooperation: Working Towards Cyber-Sovereignty*", (Published on the Henry M. Jackson School of International Studies, University of Washington website on 21 June 2016), Web URL: https://jsis.washington.edu/news/china-russia-cybersecurity-cooperation-working-towards-cyber-sovereignty/accessed on 01 May 2018
17. Kyle Russel, "*No, The U.S. Didn't Just Give Up Control Of The Internet*", (Published on the Business Insider website on 21 June 2016), Web URL: https://www.businessinsider.in/No-The-U-S-Didnt-Just-Give-Up-Control-Of-The-Internet/articleshow/32320013.cms accessed on 01 May 2018
18. List of domains blocked by the Great Firewall of China (GFW) as compiled by Hiking the GFW website, Web URL: https://hikinggfw.org/blocked_sites accessed on 01 May 2018
19. Kryptowire website, "*Kryptowire Discovers Mobile Phone Firmware that Transmitted Personally Identifiable Information (PII) Without User Consent Or Disclosure*", (Published on 15 November 2016), Web URL: http://www.kryptowire.com/adups_security_analysis.html accessed on 01 May 2018
20. Brian Krebs, "*Chinese IoT Firm Siphoned Text Messages, Call Records*", (Published on Krebs on Security website on 16 November 2016), Web URL: https://krebsonsecurity.com/2016/11/chinese-iot-firm-siphoned-text-messages-call-records/#more-36939. Accessed on 23 Jan 2017
21. Ashley Deeks, "*Tallinn 2.0 and a Chinese View on the Tallinn Process*", (Published on the Lawfare Blog website on 31 May 2015), Web URL: https://www.lawfareblog.com/tallinn-20-and-chinese-view-tallinn-process accessed on 01 May 2018
22. Website of Ministry of National Defense of the People's Republic of China, "*Central Military Commission (CMC) Departments*", Web URL: http://eng.mod.gov.cn/cmc/index.htm accessed on 01 May 2018
23. Xinhuanet website, "*China sets up joint logistics force, Xi confers flags*", (Published on 13 September 2016), Web URL: http://www.xinhuanet.com/english/2016-09/13/c_135685473.htm accessed on 01 May 2018
24. Website of Ministry of National Defense of the People's Republic of China, "*China to build world-class armed forces by mid-21st century: Xi*", (Published on 18 October 2017), Web URL: http://eng.mod.gov.cn/news/2017-10/18/content_4794926.htm accessed on 01 May 2018
25. Joel Wuthnow and Phillip C. Saunders, "*Chinese Military Reforms in the Age of Xi Jinping: Drivers, Challenges, and Implications*", (Center for the Study of Chinese Military Affairs, Institute for National Strategic Studies, National Defense University, 21 March 2017), p.23-32
26. Cristina L. Garafola, "*People's Liberation Army Reforms and Their Ramifications*", (Published on The RAND Blog website on 23 September 2016), Web URL: https://www.rand.org/blog/2016/09/pla-reforms-and-their-ramifications.html accessed on 01 May 2018
27. Anthony H Cordesman, Steven Colley and Michael Wang, "*Chinese Strategy and Military Modernization in 2015: A Comparative Analysis*", (Centre for Strategic and International Studies, 2015), p.164
28. David Finklestein, "*The General Staff Department of the Chinese People's Liberation Army : Organization, Roles and Missions*", (The People's Liberation Army as Orgnization, RAND, 2002), p.122
29. Project 2049 Institute, "*Chinese Military Modernization*", (Final Report Prepared for the US-Japan Bilateral Commission, January 2014), Appendix 6, p.2
30. Elsa Kania and John Costello, "*China's Quest for Informatization Drives PLA Reforms*",

(Published on The Diplomat Magazine website on 04 March 2017), Web URL: https://thediplomat.com/2017/03/chinas-quest-for-informatization-drives-pla-reforms/accessed on 01 May 2018

31. US China Economic and Security Review Commission, "*Report to Congress*", (US Government Publishing Office, November 2016), p.290
32. China reorients strategic military intelligence, (Jane's Intelligence Review, 2017), p3
33. David Finklestein, op. cit., p.149-151
34. GlobalSecurity.org website, "*Second [Intelligence] Department*", Web URL: https://www.globalsecurity.org/intell/world/china/pla-dept_2.htm accessed on 01 May 2018
35. Website of the Federation of American Scientists, "Second [Intelligence] Department", Web URL: https://fas.org/irp/world/china/pla/dept_2.htm accessed on 01 May 2018
36. US China Economic and Security Review Commission, *loc.cit.*
37. Mark A. Stokes, Jenny Lin and L.C. Russell Hsiao, "*The Chinese People's Liberation Army Signals Intelligence and Cyber Reconnaissance Infrastructure*", (Project 2049 Institute, 2011), p.5, 7-11
38. Mark A. Stokes and L.C. Russell Hsiao, "*Countering Chinese Cyber Operations: Opportunities and Challenges for U.S. Interests*", (Project 2049 Institute, 2012), p.3, 7-9.
39. Website of the Federation of American Scientists, "*Third Department*", Web URL: https://fas.org/irp/world/china/pla/dept_3.htm accessed on 01 May 2018
40. Website zh.wikipedia.org, "*Chinese People's Liberation Army 61398 troops*", Web URL: https://zh.wikipedia.org/wiki/>e61398 (as translated by Google Translate), accessed on 01 May 2018
41. Office of the United States Trade Representative, "*Findings of the Investigation into China's Acts, Policies, and Practices Related to Technology Transfer, Intellectual Property, and Innovation Under Section 301 of The Trade Act Of 1974*", (Executive Office of the President of the United States, 2018), p.151-176
42. US Department of Justice website, "*U.S. Charges Five Chinese Military Hackers for Cyber Espionage Against U.S. Corporations and a Labor Organization for Commercial Advantage*", (Published 19 May 2014), Web URL: https://www.justice.gov/opa/pr/us-charges-five-chinese-military-hackers-cyber-espionage-against-us-corporations-and-labor accessed on 01 May 2018
43. David Finklestein, *op. cit.*, p.167-170
44. Mark A. Stokes, Jenny Lin and L.C. Russell Hsiao, *op.cit.*, p.14-15
45. Website of the Federation of American Scientists, "*Fourth Department*" Web URL: https://fas.org/irp/world/china/pla/dept_4.htm accessed on 01 May 2018
46. Elsa Kania, "*Chinese Advances in Unmanned Systems and the Military Applications of Artificial Intelligence – the PLA's Trajectory towards Unmanned, "Intelligentized" Warfare*", (Testimony before the US – China Economic and Security Review Commission, 2017), p4
47. Bryan Krekel, Patton Adams, George Bakos, "*Occupying the Information High Ground:Chinese Capabilities for Computer Network Operations and Cyber Espionage*", (Northrop Grumman, 2012), p.45
48. David Finklestein, *op. cit.*, p.161-166
49. Mark A Stokes, Hearing on China's Military Modernization and its Implications for the United States, (The US-China Economic and Security Review Commission, 2014), p.4
50. Elsa Kania, *loc.sit.*
51. James E. Fanell and Kerry K. Gershaneck, "*White Warships, Little Blue Men, and Looming Conflict in the East China Sea – China's "Short, Sharp War" For The Senkakus*", (German-Southeast Asian Center of Excellence for Public Policy and Good Governance, CPG Online

Magazine No 4, 2017), Web URL: http://www.cpg-online.de/2017/07/01/white-warships-little-blue-men-and-looming-conflict-in-the-east-china-sea/accessed on 01 May 2018

52. Sinodefence.com website, "*Ground Forces*", Archived page accessed on Web URL:https://web.archive.org/web/20131104200224/http://www.sinodefence.com/army/default.asp accessed on 20April 2018
53. Mark A. Stokes, Jenny Lin and L.C. Russell Hsiao, *op.cit.*, p11-13
54. China Defence Today website, "*PLA Ground Force*", Web URL: https://sinodefence.wordpress.com/pla-ground-force-2016/(Updated on 01 January 2017), accessed on 01 May 2018
55. Mark A. Stokes, Jenny Lin and L.C. Russell Hsiao, *op.cit.*, p.13-14
56. Archive of Sinodedefence.com website, "*Air Forces*", Web URL: https://web.archive.org/web/20131208235620/http://www.sinodefence.com/airforce/default.asp accessed on 22 August 2016
57. Website of the Federation of American Scientists, "*PLAAF Sixth Research Institute*", Web URL: https://fas.org/irp/world/china/pla/plaaf.htm accessed on 01 May 2018
58. Archive of Chinese Military Aviation Website, "*Surveillance Aircraft*". Web URL: https://web.archive.org/web/20120601143308/http://cnair.top81.cn/y-8x_sh-5_a-50i.htm accessed on 01 May 2018
59. Sinodedefence.com website, "*Shenyang J-16*", Web URL: http://sinodefence.com/shenyang-j-16/ (Published on 15 August 2017), accessed on 01 May 2018
60. Mark A. Stokes, Jenny Lin and L.C. Russell Hsiao, *loc.cit.*
61. Sputnik News Agency website, "*Electronic Warfare Planes Join China's Navy as Military Modernization Continues*", (Published on 22 January 2018), Web URL: https://sputniknews.com/military/201801221060978414-electronic-warfare-planes-china-navy/ accessed on 01 May 2018
62. Andrew Tate, "*Images show J-15 fighter fitted with wingtip EW pods*", (Published on Jane's 360 website on 03 May 2018), Web URL: http://www.janes.com/article/79784/images-show-j-15-fighter-fitted-with-wingtip-ew-pods accessed on 20 May 2018
63. Mark A. Stokes, Jenny Lin and L.C. Russell Hsiao, *loc.cit.*
64. GlobalSecurity.org website, "*Republic of China Marine Corps*", Web URL: https://www.globalsecurity.org/military/world/taiwan/rocmc.htm accessed on 01 May 2018
65. Bill Gertz, "*Ship collisions raise specter of Chinese electronic warfare*", Asia Times, August 29, 2017
66. Larry Wortzel, "*The General Political Department and the Evolution of the Political Commissar System*", (The People's Liberation Army as Orgnization, RAND, 2002), pp.233, 239
67. Website zh.wikipedia.org, "*The People's Liberation Army 310 Unit*", Web URL: https://zh.wikipedia.org/zh-hant/-NVNl>eQ•O^ê^êŠî'n (as translated by Google Translate), accessed on 01 May 2018
68. Elsa Kania, "*The Role of PLA Base 311 in Political Warfare against Taiwan*", (Global Taiwan Brief, Global Taiwan Institute, Vol 2, Issue 7, 2017)
69. Michael Raska, "*Hybrid Warfare with Chinese Characteristics*", (Rajaratnam School of International Studies RSIS) Commentary No 262, 2015), p.2
70. Archive of Xinhuanet website, "*Chinese military conducts research on public opinion, psychological warfare, legal warfare and training*", Web URL: https://web.archive.org/web/20150930012002/http://news.xinhuanet.com/mil/2004-06/21/content_1538252.htm (as translated by Google Translate), accessed on 01 May 2018
71. Recorded Future Website, "*Recorded Future Research Concludes Chinese Ministry of State Security*

Behind APT3", (Published 17 May 2017), Web URL: https://www.recordedfuture.com/chinese-mss-behind-apt3/accessed on 01 May 2018

72. Sanjay Kumar, "*India China Space Capabilities : A Comparison*", (United Service Institution of India/Vij Books India Pvt Ltd, 2018), p.78
73. Sanjay Kumar, *op.cit.*, p.68
74. *Ibid*
75. Sanjay Kumar, *op.cit.*, p.79
76. Sanjay Kumar, *op.cit.*, p.82
77. GlobalSecurity.org website, "*SC-19 ASAT*", Web URL: https://www.globalsecurity.org/space/world/china/sc-19-asat.htm accessed on 01 May 2018
78. GlobalSecurity.org website, "*Kunpeng-7/DN-2*", Web URL: https://www.globalsecurity.org/space/world/china/kunpeng-7.htm accessed on 01 May 2018
79. Sanjay Kumar, "*China's White Paper on Space : 2016*", (Published on The United Service Institution of India Blog on 12 January 2017), Web URL: http://usiblog.in/2017/01/chinas-white-paper-on-space-2016/accessed on 01 May 2018
80. GlobalSecurity.org website, "*Chinese Demonstrates Satellite to Satellite Rendezvous, Inspection Test Bed for Duel Purpose Military Technology*", Web URL: https://www.globalsecurity.org/space/world/china/sj-12.htm accessed on 01 May 2018
81. GlobalSecurity.org website, "*Chinese Anti-Satellite [ASAT] Capabilities*", Web URL: https://www.globalsecurity.org/space/world/china/asat.htm accessed on 01 May 2018
82. Ian Easton, "*The Great Game in Space China's Evolving ASAT Weapons Programs and Their Implications for Future U.S. Strategy*", (Project 2049 Institute, 2014), Appendix 6, p.2
83. Sanjay Kumar, *op.cit.*,p.83-85
84. *Ibid*, p.86
85. Website of Ministry of National Defense of the People's Republic of China, "*Xi calls new PLA branch a key pillar*", (Published on 31 August 2016), Web URL: http://eng.mod.gov.cn/Opinion/2016-08/31/content_4728709.htm accessed on 01 May 2018
86. Xinhuanet website, "*China reshuffles 84 corps-level military units*", (Published on 18 April 2017), Web URL: http://www.xinhuanet.com/english/2017-04/18/c_136218258.htm accessed on 01 May 2018
87. John Costello, "*The Strategic Support Force: China's Information Warfare Service*", (The Jamestown Foundation, China Brief, Volume 16, Issues 3 and 19, 2016)
88. John Costello, "*China's Strategic Support Force: A Force for a New Era*", Testimony to the U.S.-China Economic and Security Review Commission, February 15, 2018
89. Todd Harrison, Kaitlyn Johnson, Thomas G Roberts, "*Space Threat Assessment 2018*", (Center for Strategic and International Studies, 2018), p.7
90. China Defence Today website "*PLA Strategic Support Force*", (Updated on 01 January 2017), Web URL: https://sinodefence.wordpress.com/pla-strategic-support-force/accessed on 01 May 2018
91. *Ibid*
92. Elsa Kania, "*PLA Strategic Support Force: The 'Information Umbrella' for China's Military*", (Published on The Diplomat Magazine website on 01 April 2017), Web URL: https://thediplomat.com/2017/04/pla-strategic-support-force-the-information-umbrella-for-chinas-military/accessed on 01 May 2018
93. Sina Military Website, "*Experts say that the strategic support force's independent military concept is ahead of the US military*", (Published on 08 January 2016), Web URL: http://mil.news.sina.com.cn/china/2016-01-08/doc-ifxnkkuy7732953.shtml (as translated by

Google Translate), accessed on 01 May 2018

94. Joel Wuthnow and Phillip C. Saunders, "*Chinese Military Reforms in the Age of Xi Jinping : Drivers, Challenges, and Implications*", (US National Defense University Press, 2017), p 12
95. *China reorients strategic military intelligence*, (Jane's Intelligence Review, 2017)
96. Elsa B. Kania and John K. Costello, "*The Strategic Support Force and the Future of Chinese Information Operations*", (The Cyber Defence Review, Volume 3, Number 1, Spring 2018), pp.111-113
97. Elsa Kania, ""Unmanned, Intangible, Silent Warfare" – New Threats and Options for Taiwan", (Global Taiwan Brief, Global Taiwan Institute, Vol 3, Issue 3, 2018)
98. Chinese Military Aviation Blog, "*UAV/UCAV II*", Web URL: https://chinese-military-aviation.blogspot.in/p/uavucav-ii.html accessed on 01 May 2018
99. Elsa B Kania, "*The Role of PLA Base 311 in Political Warfare against Taiwan*", *loc.sit.*
100. Captain James Fanell (USN, Retd), "*China's Global Naval Strategy and Expanding Force Structure: Pathway to Hegemony*", (Testimony to the U.S. House of Representatives, Permanent Select Committee on Intelligence, 2018), p24
101. James E. Fanell and Kerry K. Gershaneck, "*White Warships, Little Blue Men, and Looming Conflict in the East China Sea – China's "Short, Sharp War*" *op.cit*, p.29
102. Sputnik News Agency website, *"Electronic Warfare Planes Join China's Navy as Military Modernization Continues*", (Published on 19 January 2016), Web URL: https://sputniknews.com/asia/201601191033349605-china-strategic-support-forces/accessed on 01 May 2018
103. Peter Mattis and Elsa Kania, "*Modernizing Military Intelligence: Playing Catchup –II*", (Published on Ocnus.net website on 21 December 2016), Web URL: http://www.ocnus.net/artman2/publish/Dysfunctions_2/Modernizing-Military-Intelligence-Playing-Catchup—II.shtml acessed on 01 May 2018
104. Military Mandarin website, "*GHQ Vocabulary List*", Web URL: http://www.zhanyixue.org/ghq-vocab-list.html accessed on 01 May 2018

CHAPTER 8

Conclusions

While consolidation of all information related capabilities might seem lucrative, the information domain is all pervasive and consolidating all capabilities into a monolith may not be the best idea operationally. It would be appropriate to summarize the important lessons we have learnt from IW theory and analysis of actual organizations.

Separation of Offensive and Defensive Capabilities

Unlike other battlespaces where aircrafts and ships may engage other aircrafts and ships, IW systems don't attack similar IW systems. Therefore the battle is not Jammer vs Jammer; instead it is Jammer vs Receiver. Similarly it is Cyber Warfare vs Networks, Psy Warfare vs Cognition and Espionage vs Information Security. It is the converse while defending. The same capabilities cannot be used to attack as well as defend. Equipment and tactics are completely different and force structures have to be tailor made for either offensive or defensive role. It can never be both. This is exactly similar to the land battlespace.

Typically a collection of ESM, ECM, CNA and SIGINT/CNE units will comprise the offensive formations (viz the Network System Department of China's SSF) while the communication, networking and CND units will comprise the defensive formations (viz the US Cyber Command). It serves no purpose to mix up the two, rather it may be counterproductive. It is logical to have two separate forces, one for attacking and one for defending.

As we shall see in the points below, it is also counterproductive to combine each and every offensive IW capability. It has to be a narrow and careful selection.

Communications and Networking are Integral to IW

While we have defined only Psy Warfare, Intelligence, EW and Cyber Warfare as core IW capabilities, what has emerged from our analysis is that defensive IW operations are equally important and they are inseparable from the underlying communication networks. The revised US IW Doctrine rightly includes DoD Information Networks (DoDIN) as an IRC and the US Cyber Command is essentially all about consolidating the communications and networking capability in a manner that it guarantees information assurance despite attacks from the core IW capabilities we have defined in this book. "Communication Networks" should therefore be treated as yet another core IW capability because it is here that own information cycles play out, particularly in the electronic terrain. No wars or battles can be won if own information cycles cannot survive enemy IW assault. This calls for a major doctrinal overhaul but not necessarily a big change to force structures because command and control networks already exist in every military. If communication networks are not treated as a core IW capability then the underlying doctrine loses the defensive capability.

SIGINT – Cyber Warfare Convergence

SIGINT and EW (the ESM portion) were identical (or complementary) in the pre-networking era when ESM entities of EW would pick up the same radio, microwave and radar signals which the COMINT and ELINT systems did. The only difference between SIGINT and ESM was that SIGINT systems were often deployed at selected vantage points and collected intelligence by monitoring peacetime communications whereas EW (ESM) systems were mobile and grouped with fighting echelons to be deployed during operations and in close proximity to the enemy for monitoring communications that came up in the tactical battle area (TBA). Theoretically this meant that SIGINT would provide strategic intelligence and EW will yield tactical intelligence (often called Tac SIGINT) including the Electronic Order of Battle (E-ORBAT). However, it was often the case during operations that the same emitters were intercepted and reported by both EW and SIGINT units, possibly because SIGINT units also found the new emitters that opened up during war to be more lucrative targets. While the duplicity caused concern, the fact that the raison d'être for the ESM portion of EW was to identify targets for jamming (or engagement by kinetic means) and that modern integrated EW systems could automate the entire process was often lost on lay observers. All this has dramatically changed in past two decades.

The scope of SIGINT has completely transformed and significantly expanded in the ICT era. New forms of communication and transmission have matured and became commonplace. Networking has replaced point-to-point communications and digital microprocessors have taken over the switching functionalities. Voice, data, video and machine to machine communications are no longer distinct signals but a common stream of digital packets flowing on data networks which in turn are a mix of wireless and wired technologies deployed across land, sea, air and space. Likewise it is not just radars which radiate electromagnetic energy. A plethora of devices ranging from household remotes to driverless cars emit some form of electromagnetic energy to support their functionalities. SIGINT had to embrace all these to stay relevant and it did. COMINT expanded to include cell phones, satellite communications, and even tapping of wired or wireless computer networks and fiber optic cables. ELINT likewise expanded beyond radars to cover many different types of signals like telemetry, beacons and missile control; and branched off into sub-specialties like MASINT and FISINT.

Modern SIGINT systems are accordingly deployed on ground, underwater, on UAVs and in space. Bandwidths, bit rates and data volumes which used to be expressed in kilo (10^3) binary prefix in the early 90s became mega (10^6) a decade ago and are now in the Giga (10^9) and Tera (10^{12}) range necessitating use of computers for data interception and traffic analysis. Also the boundary between computers and communications has completely vanished. In today's era of mobile computing and wireless networking – most devices (viz smart phones) are both a trans-receiver as well as a computer. Interception has morphed into exploits which have elements of both SIGINT and CNE stitched together. One of the simplest examples is hacking of a WiFi router. This offensive flavor does not however convert SIGINT into EW – it simply shows that SIGINT blends with CNE and vice versa. Also the old definitions restricting SIGINT to passive intelligence gathering are no longer accurate and require to be changed. Another major area of congruence between SIGINT and Cyber Warfare is cryptanalysis of encrypted communications which relies heavily on computer facilitated cryptographic attacks. It would therefore not be wrong to say that SIGINT, Cryptology and Cyber Warfare (CNE) have converged and become inseparable. The NSA/CSS and 3PLA are living proof of this. The US doctrine abandoning the term CNE and now only categorizing Cyberspace Operations into Offensive and Defensive is also possibly a result of this realization. Since CNE generates targets for CNA, it also makes sense to link the CNA operations with SIGINT, particularly at the strategic level.

SIGINT – EW Divergence

Till the 80s or so (depending on force modernization) many military radio sets and point to point multi-channel microwave links did not use proprietary waveforms, encryption and ECCM techniques. It was therefore often possible for EW systems to record enemy voice communications or extract data from intercepted multi-channel signals. The situation has dramatically changed since then. Most modern CNR sets (and microwave links) use sophisticated encryption and ECCM techniques like frequency hopping and spread spectrum. They also transmit proprietary digital waveforms instead of standard modulated signals. Interception, demodulation, demultiplexing and decryption of such communication signals are an extremely arduous task. Even location fixing of a frequency hopping/spread spectrum radio emitter requires sophisticated ESM equipment. Many old timers however still believe that the main purpose of EW is to eavesdrop into battlefield conversations of enemy commanders. That era is dead. Whereas earlier both EW and SIGINT units could extract voice and data from intercepts, it is now mostly SIGINT units which are still able to do so because they tap into civil communication networks that use laid down frequencies, universal protocols and weaker encryption. Sometimes the service providers themselves facilitate interception as part of legal obligation. In the modern battlefield, the ESM portion of EW is now mostly about generating targets for jamming, facilitate engagement by kinetic means and generate situational awareness data through location fixing. If one were to summarize the role of EW systems on the battlefield it would be as under:

- Carry out target specific jamming of the electromagnetic spectrum leading to denial of command and control, navigation and PNT as also non-kinetic denial of space based assets
- Carry out target specific jamming (or deception) of weapons and platforms that rely on use of electromagnetic spectrum for activation, detection or targeting thereby disabling or misguiding them. Counter RCIED is an example. Jamming, spoofing of radio/satellite navigation systems and some forms of ECM attacks designed to trick radars are yet another example. EW suites to protect ships and aircrafts against missiles are another major application
- Integration with kinetic means either as an alternative to impose delay and destruction or be employed in tandem to enhance the degradation,

accuracy and lethality of the collusive attack (this includes Directed Energy Weapons)

- Facilitate continuous generation of enemy ORBAT through location fixing and tracking of military emitters in conjunction with GIS and ISR capabilities, thereby providing battlefield transparency and early warning
- Provide Tac SIGINT by geo-locating and characterizing intercepted signals (and possibly also use it to segregate friendly emitters, though that is primarily a spectrum coordination issue)

Amongst the five above, it is the first three tasks that are the primary ones, and though facilitated by ESM, they have very little in common with SIGINT or Cyber Warfare. Yet the myth that EW, SIGINT and Cyber Warfare are converging and must be unified remains alive.

EW – Cyber Warfare Convergence

Another widely held myth is that EW jammers can act as carriers of malicious payloads to be injected into enemy networks and platforms. This needs amplification. ECM systems are designed to transmit jamming signals and not to communicate with receivers. ECM systems may retransmit, mimic or modify signals to deceive enemy receivers (particularly radars); however this is a standard EW technique and does not tantamount to injection of any malicious payloads (ie cyber warfare). Modification of transmission medium by using chaff is likewise EW and not cyber warfare. Spoofing of GPS signals is similarly an EW technique and not cyber warfare. If we exclude these types of EW attacks, and consider whether a jammer can actually insert a malicious code (software) into encrypted data streams, and whether such payloads can then be executed on enemy's computational devices in a verifiable way is a very questionable assumption. What SIGINT and Cyber Warfare may achieve with commercial civil communications is difficult to replicate in the TBA.

To summarize, EW has little in common with Cyber Warfare, particularly in the TBA. EW-Cyber Warfare unification is at best possible at the strategic level (viz ASAT operations which may use a combination of ECM and CNE/CNA) and not at the tactical level. The Chinese SSF follows this rationale.

Space-IW Convergence

Space is mostly an unmanned arena and entities deployed in this battlespace must per force use wireless links, electronic hardware, and software for command, control, telemetry and relaying back data. This makes space based assets immensely vulnerable to EW and Cyber Warfare. In fact the non-destructive, reusable, simple and inexpensive nature of these IW techniques combined with the advantage of deniability makes them the preferred method of attack vis-à-vis kinetic means.

What also merits attention is that space based assets lie at the heart of not just long haul communications but also ISR, PNT and BMD capabilities which are vital in the modern battlefield. Any substantial degradation of these capabilities has the potential of changing the entire operational dynamics and severely downgrade the adversary's warfighting potential. Expeditionary forces like the US and offensive formations which fight deep within enemy territory are the most vulnerable.

Subdivisions of EW

Tactical EW resources are most effective when grouped directly with the fighting formations (field formations/carrier groups/air commands) and they are the primary capability which integrates IW with kinetic means. These should ideally not be centralized except where paucity of equipment requires a mission based allocation. Integrated EW systems help optimize the resources and are also more suitable for integration with kinetic means as compared to standalone systems.

Strategic EW (mostly ASAT and SEAD) assets need to be treated separately and can be grouped at the theatre level to enable planned application and concentration of firepower.

EW assets integrated with platforms for self-protection and stealth (viz EW suites of aircrafts and ships) are a third distinct category.

Both the US and Chinese IW organizations reflect this broad philosophy.

Integration with Kinetic Means

For optimum effect, IW must be integrated with kinetic means. However, combining kinetic hard-kill and IW soft-kill weapons under a single headquarter is not the best way to achieve integration. The artillery model of grouping integral resources abinitio and augmented by theatre resources on an as required basis is much more appropriate.

Intelligence Overlap

Intelligence is a major IW capability that covers the physical terrain. It therefore supports Psy Warfare through HUMINT, and Cyber Warfare and EW capabilities through sabotage. Above all intelligence consolidates the acquisition operations across all terrains and this is vital to corroboration and coordination. This however also results in overlaps with the other capabilities. SIGINT for example is a major component of intelligence yet also shares big synergy with Cyber Warfare. Any attempts to completely merge SIGINT with Cyber Warfare will not only seriously cripple the intelligence gathering capability but also break the vital link between the electronic, physical and cognitive IW operations. It is however also not possible to detach SIGINT from Cyber Warfare.

An overlap and duplication between intelligence operations and the other core capabilities must therefore be tolerated. If sufficiently trained, the personnel can also be exchanged amongst the technical capabilities. The US NSA/CSS is an example.

Similar overlap of intelligence functionalities must be tolerated when it comes to military and non-military organizations. Consolidating the diverse intelligence requirements under a common organization is wasteful and illogical.

Psy Warfare

The realms of cognitive and electronic information are moving closer (viz. social media) but cognitive operations cannot be detached from events on ground. In the battlefield Psy Warfare is best carried out by troops which are on ground and in contact with the enemy/local civilian population but during peace Internet and Social Media play an increasingly important role in shaping perceptions. Counter insurgency is a mix of both. Psy Warfare therefore needs to integrate both with Cyber Warfare and HUMINT to be fully effective. It would therefore be important to group some Psy Warfare capability with Cyber Warfare while leaving bulk of them with the field formations. The Chinese SSF seems to follow this.

It also merits attention that ground forces invariably act as the lead agency for Psy Warfare due to their physical proximity with local inhabitants or enemy combatants. The Navy and Air Force play a supporting role.

Increasing Importance of Electromagnetic Spectrum

As more and more machines join the battlefield and as network-centricity increases, the role of secure wireless for command, control, PNT, ISR and Identification, Friend or Foe (IFF) expands rapidly. The role of spectrum management (defensive) and EW is increasing exponentially. This becomes even more pronounced and critical in case of expeditionary forces and offensive formations fighting deep inside enemy territory which is why some analysts have started identifying electromagnetic spectrum as a unique battlespace over and beyond information. We can avoid this complication if we are able to treat spectrum as a part of the electronic information terrain and treat spectrum management and ECCM as defensive operations conducted on this piece of ground whereas ESM and ECM are the offensive ones.

Application Specific Weapon Systems

EW equipment is extremely application specific. The EW suite integrated into an aircraft cannot jam RCIEDs, the RCIED jammer in turn cannot disrupt V/UHF Tac CNR and a V/UHF jammer cannot jam a satellite or act like a GPS jammer. Same applies to SIGINT and ESM. The hardware, waveforms, antennae, control software are all completely different. Careful planning is required to prioritize and acquire systems to cover the entire spectrum of operations and targets.

Weapons used for Cyber Warfare are similarly target (hardware and software) specific. Once again detailed planning is required to identify, prioritize and acquire weapon systems which cover the entire spectrum of targets. Systems required for defense (CND) are again completely distinct from the offensive ones. The arsenal must have all the required offensive and defensive weapon systems if success is to be achieved.

Integration Amongst Services

While theoretically merger of all IW capability of Army, Navy and Air Force appears lucrative, there are many challenges. Firstly, the defensive IW capabilities can certainly be merged and such integration of C4ISR capability can act as a tremendous force multiplier and reinforce network-centricity. However unified theatre commands are a prerequisite for any such tri-services integration. Attempting to integrate the defensive IW capabilities (communications, networks, CND) without a unification of command will only deprive commanders of the intimate support that they require for command and control.

Cryptologic support and management of spectrum have to be integrated regardless of unification.

When it comes to offensive operations, it again requires careful deliberation. In case of EW it does not make much sense to integrate ground based EW units with air force and naval EW assets or to combine them into an integrated force. The employment concepts, operations and tactics are completely different and the logistic and administrative complexities involved in functioning of such a unified force are nightmarish. As already brought out, these are best left integrated with the respective fighting formations. It would be much more sensible to have some UAV/helicopter based entities as part of the ground based EW systems. The Air Force/Navy may likewise have some specialized ground EW equipment like manpack EW systems/Counter RCIED systems for special operations. Joint organizations similar to what exist for air-defense are however required to avoid fratricide.

Coming to Cyber Warfare, it is possible for SIGINT/CNE/CNA operations of all the services to be combined and they may also benefit from the enhanced range and additional options which such a merger can leverage (viz. naval forces are best suited to target trans-oceanic submarine cables whereas air force is much more effective for collection of SIGINT from depth areas).

Psy Warfare capability is mostly led by ground forces with Air Force and Navy providing assets. In both cases, unified theatre commands are extremely desirable for optimum synergy.

Good and Bad Combinations

We are now in a position to identify which combination of capabilities is harmonious and maybe combined into force structures and which ones are not amenable to be grouped. Such analysis also helps in formulation of operational doctrine and application of IW capabilities in synchronicity with kinetic weapons and capabilities in the other battlespaces. The combinations are explored below:

- **Good Combinations.** The following combination of capabilities work very well:
  - *Communications + Networking + CND.* The US Cyber Command is an example. This is an excellent combination for defensive operations and force structures.
  - *SIGINT + CNE + CNA + Sabotage.* This combination maximizes the electronic acquisition and attack capabilities. The US NSA/CSS is an example. Sabotage capabilities maybe built into the organization or mechanisms evolved to borrow these from intelligence organizations.
  - *Psy Warfare + HUMINT + SIGINT + CNE.* This combination maximizes effects in the cognitive realm wherein it can leverage both the physical as well as electronic terrain to influence hearts and minds. It may however not be possible to create such a diverse organization and the integration maybe achieved through procedures and joint organizations.
  - *EW (Tactical) + Kinetic Weapons.* This combination is vital to winning future wars. The integration is not only at the procedural level but also hardware and software.
  - *EW (Strategic) + CNE + CNA + ASAT Kinetic Weapons (Counter Space).* Non-lethal strategic ASAT EW systems and destructive counter-space kinetic weapons are complementary and can be applied in unison or as an alternative to each other. CNE and CNA can assist by targeting the ground based command and control setup but this is a less guaranteed attack vis-à-vis EW. China's SSF is evolving towards such a force structure.
  - *All Acquisition Operations (All Intelligence Disciplines like HUMINT, SIGINT, MASINT, FISINT etc + CNE).* Consolidation and fusion

of all acquisition functions is vital to provide timely and accurate intelligence. All intelligence organizations must follow this rule.

- *SIGINT + Cyber Warfare.* This is again a very logical combination. It however remains a dilemma as to whether SIGINT should be clubbed with intelligence or Cyber Warfare. The solution lies in logical splitting of tactical and strategic resources as also evolving procedures and joint organizations. The US NSA/CSS and INSCOM are an example.

- **Bad Combinations.** The following combination of capabilities do not work very well:
  - *CNE + CNA + CND.* Offensive and defensive operations cannot be combined under a single formation.
  - *EW + Cyber Warfare + Psy Warfare.* The three capabilities operate in different terrains and such all-encompassing merger of all offensive and defensive capabilities into a unified force is illogical, wasteful and counterproductive.
  - *EW + SIGINT.* Though they share synergy in terms of acquisition, the disciplines have diverged. Combining them has more disadvantages then advantages.
  - *EW + CNA (Tactical).* EW systems have a completely different role and design and must be kept distinct from Cyber Warfare. EW – Cyber Warfare integration is theoretically possible if systems (hardware and software) are available which can mimic an enemy switching node, terminal or encrypted transmission on a sustained basis. EW can then act as a vector to allow infiltration into enemy tactical data networks and launch computational attacks. Achieving this practically is very difficult. Permanent force structures cannot be created based on such hypothetical assumptions.
  - *Intelligence + ESM + CNE.* Despite being acquisition operations, these cannot be merged. The main role of ESM and CNE is to facilitate attacks and provide integrated real time support to offensive operations. Equipment and underlying tactics are different. SIGINT meets this requirement when it comes to intelligence.

- *Army + Navy + AF – Unified Command.* Consolidation of IW capabilities of all branches without creation of a unified command structure is counterproductive.

Conclusion

It is not surprising that there is a race amongst advanced nations to occupy dominating heights of the information battlespace. These efforts involve many challenges like evolving new doctrines, acquiring IW weapons and systems, their integration amongst themselves and with kinetic means, creating organizations and leadership, as well as creating trained manpower to handle these specialist tasks.

We have also seen that information is a unique and diverse battlespace. It poses many doctrinal dilemmas and the underlying capabilities are often strange bedfellows. We may draw an analogy with a chef who wants to create a new dish. The chef firstly needs to be aware of what spices combine well and which ones don't. He secondly needs to identify the ones which will yield the required flavor. Lastly he has to be clear about the quantity in which each of the selected ingredients has to be used. Failure in any of the three steps will result in a bad dish.

IW doctrines and force structures are similar. Firstly we have to be clear which capabilities combine and which ones don't. Secondly we need to be clear of what is it that we wish to create to meet our strategic objectives. Is it a defensive formation like the US Cyber Command or an offensive formation like the Chinese SSF? It can't be both simultaneously. Lastly, we have to identify the correct mix of complementary capabilities to create a force structure which will fulfill the desired role. Most IW doctrines fail in meeting all these criteria and instead just try to catalogue all the capabilities without going into specifics.

If we try to analyze the interaction amongst the five battle spaces of land, sea, air, space and information, we find that while each one plays an enabling role for the others, it is information which unifies them. Operations to defend own information systems against enemy attacks enable fighting a coordinated battle in the other four battle spaces whereas destroying the adversary's information systems force him to fight isolated, uncoordinated battles in an information vacuum. Due to its overarching nature – the winner in the information battlespace is much more likely to win in the other battle spaces as well. This however also implies that information as a battlespace cannot be treated in isolation. The mission and composition of the Armed Forces i.e. whether they are meant for force projection or for a primarily defensive role dictates the IW force composition. The nature of asymmetry amongst the opposing forces is also an important factor. This is the main reason why the US Cyber Command is a defensive formation while the Chinese SSF is an offensive formation. Terrain and enemy tactics also play a role in deciding which IW capability needs to be given priority.

It is also evident that amongst the three types of information (viz. cognitive, physical and electronic); it is the electronic terrain which has become predominant in the modern network centric battlefield. The side which is able to ensure superior flow of electronic information will not only achieve optimum synchronization amongst various kinetic means, it will also enjoy superior Command & Control, ISR and PNT capabilities as also a greater ability to spring (or counter) surprises. Vice versa, the side which loses the defensive electronic battle is doomed unless it possesses overwhelmingly superior force levels, weaponry and armament or is able to execute some extraordinary victories in the cognitive or physical realms of information. Winning the war despite losing the electronic battle is highly improbable.

This does not imply that cognitive and physical realms of the information battlespace have become insignificant; however the role played by them diminishes in a short intense war and increases in case of a prolonged campaign or a standoff. Modern conventional wars are increasingly likely to be swift and brutal with victory decided in the first few days and through offensive actions. Superiority in the electronic realm is therefore a much bigger decisive factor in the crucial early stages of war and ranks at par with other vital determinants like force levels, speed of mobilization and combat readiness.

Finally political factors and the type of governance also play a major role in determining the IW force structures. This is because of the over-arching and multi-disciplinary nature of information and the dual civil-military nature of IW capabilities. The role of the military and the trust it enjoys with the national leadership will determine the manner in which capabilities are assigned and integrated at the highest level. The IW force structures suitable for a democratic country will not be suitable for a nation which has a single party rule. Capitalist and communist regimes will similarly have different force structures. The PLA SSF may not appear to be perfect in theory but that is because it also must be politically correct.

There is no set piece doctrine or force structures which can be applied universally to every scenario. Each country has to identify its own unique flavor depending on the geostrategic and political compulsions, the threats it faces and the division of responsibilities amongst various organs of national power. A holistic analysis of all these factors combined with the basic principals we have seen in this book will thereafter allow creation of a proper doctrine as well as the optimum IW force structures.

EPILOGUE

IW in the Indian Context

Epilogue
IW in the Indian Context

Indian Armed Forces have been involved in four major wars since independence. These were the Indo-Pak wars of 1947, 1965, 1971 and the Sino-Indian war of 1962. All these were pre-NCW, pre-IW era conflicts. We may extrapolate to derive some lessons, but they do not really provide any big insights into modern IW force structures and the resultant operational doctrine.

The brutal conflict in Sri Lanka (1987-90) between the Indian Peace Keeping Force (IPKF) and the Liberation Tigers of Tamil Eelam (LTTE) was also pre-IW but saw EW capability coming into play. One of the heaviest casualties suffered by the IPKF during this conflict was during the Jaffna University helidrop in which Indian soldiers landed into a well planned LTTE ambush and 35 of them laid down their life[1]. Military experts have ascribed various reasons for this debacle, but most important of them actually was that the LTTE had intercepted IPKF VHF radio communications and were already aware of the helidrop and waiting to kill. The second most important reason was that communications between the IPKF headquarters, the helicopter pilots and the battalions which were heli-dropped was poor. Though nowhere near the same magnitude as the Iraqi annihilation during Operation Desert Storm in 1991 – the debacle underscores the fact that defensive IW capability (i.e. secure and fail-proof command and control communications) is extremely crucial to victory and its absence completely nullifies any superiority in firepower and maneuver. Dr Carlo Kopp has written in detail about the Electronic Battle during operation desert storm in Iraq and it is a very instructive read[2]. This period (1980s) saw the raising of ground based EW units in the Indian Army and these constitute the backbone of the tactical IW force structure of the ground forces.

The limited but intense war in Kargil against Pakistan (Operation Vijay,1999) was fought in the post NCW era but the Internet (IW) age had not yet dawned. The operations did contain several important lessons from the IW perspective. The most important of these was about the influence of media in shaping perceptions. This was the first televised war fought by India and media acted as a tremendous force multiplier. The advent of Internet and

Social Media has complicated the dynamics since those early days. These new vectors have now superseded television and radio and nobody has intimate control over information anymore. Maintaining the nation's will to fight and influencing perceptions has become contingent upon the ability to attack and defend on the Internet. Synergistic conduct of IW operations by all the elements of national power has accordingly become much more complex. The second most important lesson learnt from the extremely rugged high altitude battles in Kargil was the need for terrain specific equipment. This applies to IW as well, particularly the tactical EW systems. IW equipment and force structures must cater to terrain just like any other capability. Thirdly, the Kargil conflict was also India's first encounter with navigational warfare when the US reportedly denied GPS data to India, spurring India to develop its own indigenous capability[3]. Lastly, it was the EW/SIGINT capability which intercepted a call from Beijing between Pakistani General Pervez Musharraf and his Chief of the General Staff Lt Gen Mohammad Aziz Khan which defeated the Pakistani grand strategy of claiming that it was mujahideen fighters who had occupied Indian territory and not its regular Army[4]. The age of cyber warfare had not yet dawned.

The role of IW has exponentially grown since then. The period following operation Desert Storm was when India realized the urgency to develop indigenous EW capability. The Samyukta (a Hindi word for integrated) indigenous EW program was initiated in 1994 under the visionary leadership of the then defense hierarchy, who rightly identified such systems as an absolute necessity. Samyukta was intended to be a long term Integrated EW Program (IEWP) on similar lines as the Integrated Guided Missile Development Program (IGMDP) and was steered by similar high level empowered committees. However, while the IGMDP became a great success; the IEWP faltered and stalled despite the fact that the first block was handed over to the army with great fanfare in a ceremony presided over by the then Indian president Dr APJ Abdul Kalam and the then Defense Minister Mr. George Fernandes in 2004[5]. No major indigenous integrated EW systems for conventional warfare got developed after this. Two lessons can be drawn from this attempt. Firstly, IW technologies are one of the most closely guarded secrets and developing and sustaining state-of-art IW systems is even more challenging then developing an indigenous missile capability. Secondly, developing indigenous IW equipment (both hardware and software) requires sustained long term effort and it must be steered as a special program instead of as any other conventional weapon system. If Indian Armed Forces have to break free from obsolete tactics and strategy of the previous century, then it must develop such indigenous capability regardless of the cost and effort. All modern Armed Forces are striving to achieve this ability and it is oft heard amongst "crows[6]" that an imported EW system is already compromised. Victory in the future battlefields dominated by intelligent systems and machines will depend almost entirely on the electronic battle instead of tactical acumen and valour. Indigenous EW capability is vital to victory.

The next moot point is that the Indian political hierarchy lacks a long term military vision. Even the armed forces (particularly the ground forces), have become excessively focused on counter-insurgency and while this has resulted in the troops being battle hardened and combat ready to an extent, it has also led to complaisance about the changing nature of the modern battlefield and its dynamics. The belief amongst political leadership that the last major war was fought many decades ago (or can be deterred through diplomacy) is extremely fatalistic. Kargil was a complete surprise and India again nearly went to war with Pakistan after the terrorist attacks on the Indian Parliament in 2001. The 2008 Mumbai terrorist attack came out of the blue. The 2016 suicide attack on the Uri Army base which led to India launching a surgical strike on Pakistan was again a surprise. The Doklam standoff with China in 2017 was yet again very sudden and could well have escalated. It is therefore only a matter of time before adversaries become technologically confident enough to also test India in a high-tech war. State-of-art IW capability and appropriate force structures is a must if such misadventures are to be detected, deterred and thwarted. This also includes developing state-of-art technical intelligence capability and extending secure communication bandwidth to the forward edge of the battlefield.

It was in the year 2005 that Indian Armed Forces came out with their first Joint IW Doctrine. It has not been revised since then and is obsolete. The Indian Army issued an IW doctrine for the Indian Army in 2010. Though not obsolete, it now needs to be urgently reviewed – more so because the Indian Army is planning to downsize and modernize on similar lines as other major military powers[7]. Substantial piecemeal augmentation of IW capability has taken place over time but a clear vision and systematic long term planning to fully integrate IW capability into warfighting doctrine and force structure is the need of the hour. Every other country is doing so.

Even in case of insurgencies and terrorism, the nature of the beast has changed. Today, it is operations in the information battlespace which facilitate command and control, indoctrination, finance, as well as recruitment and training. Remaining oblivious to these new dynamics is self-defeating.

Clearly, IW force structures and doctrine are dictated by geostrategic perceptions, adversaries, terrain, equipment, indigenous capability as well as domestic considerations. How then should India go about it?

Terms of Reference in the Indian Context

In military parlance "terms of reference" is used to denote compulsions which must be taken into account while formulating a plan. Such terms of reference can be further subdivided into two categories – firstly the universal ones that are dictated by first principles; and secondly those which are specific to the ground realities ie the India specific considerations.

Universal Terms of Reference

We have already discussed the universal terms of reference in depth and the key ones are summarized in Chapter 8. These are applicable to any IW force structure and doctrine regardless of any other considerations. For example, a sound IW force structure will tolerate the overlap between the intelligence (acquisition) capability with psy warfare, EW and cyber warfare instead of trying to merge them. Similarly a sound IW force structure will cater to both attack and defense instead of merely focusing on the offensive IW capability. We will not repeat all of them again.

India Specific Terms of Reference

No two countries are similar and we have seen how the US Cyber Command and China's SSF are completely opposite and designed to counter each other in the information battlespace. India is different from both US and China and it makes sense to have a force structure and doctrine that is tailor made to suit her own geostrategic objectives and compulsions. Let us look at some of the key India specific terms of reference.

- **Grand Strategy**. India lies in a hostile neighborhood. It shares long unsettled borders with China and Pakistan with a history of wars and skirmishes with both of them. Potential exists for any of these border incidents or terrorist attacks to suddenly and unpredictably escalate into war as happened in 1999 and almost again in 2001. China as we have seen is rapidly modernizing its capability to dominate the space and information battlespaces. Pakistan has likewise upped the ante in Kashmir and its long term strategy of cross border terrorism is now fully supported by well planned IW operations. The Indian IW force structure must therefore be capable of addressing both the terrorist threat as well as the threat of a high-tech information and network centric war. It must be noted that the two require different types of equipment, tactics and integration (force structures). The CI capability has to be built around winning the battle for perception (psy warfare) whereas the conventional capability has to be built around dominating the electromagnetic spectrum and ensuring survivability of own command and control networks. When it comes to cyber warfare the focus of operations in CI is to neutralize the hostile terrorist organization's command and control, funding, indoctrination and propaganda and integrate with psy warfare whereas the focus of conventional capability is to target the enemy critical information infrastructure while protecting our own. Intelligence operations likewise differ in scope and conduct during CI and war. The situation is exactly similar to how CI Forces (Rashtriya Rifles and Assam Rifles) share some commonality with conventional forces but are unique and distinct in their role, organization, equipment holding and tactics. The Indian IW capability must therefore suitably support both these force structures and cater to both CI and conventional war.

- **Existing Force Structures.** India does not have unified theatre commands though the idea has been under consideration for a long time. Neither the US nor the Chinese doctrines and force structures can therefore be directly applied to the Indian scenario. This however does not imply that the IW capability cannot be optimized. We have already seen that Army plays the lead role in psy warfare with the Air Force and Navy supporting it. This can be implemented even without unified theatre commands through joint staff organizations. Likewise it does not make much sense for each service to have independent offensive cyber warfare and SIGINT capability and they need to be consolidated under a single agency like the US NSA which functions directly under the DoD. In India's case the HQ Integrated Defence Staff (IDS) is the closest equivalent. When it comes to EW, we have seen that it is best integrated at the tactical level and the requirements of the Army, Navy, Air Force and Coast Guard are distinct and unique. Even in a unified command scenario, it is best to integrate the EW capability only through procedures and joint staff organizations rather than create any combined forces comprising of aircrafts, ships and ground based systems. The joint EW centers can function on similar lines as air space control and air support organizations and they must cater to joint SEAD operations, joint amphibious and airborne operations as well as preventing electronic fratricide. The only exception to this rule is EW against enemy space based assets which must be consolidated. When it comes to the intelligence capability, we have seen that it is crucial to real time planning and conduct of operations and the requirements are service specific. A robust intelligence sharing and tasking mechanism is however essential and must be achieved via joint staff organizations and unified databases. One essential requirement is that all the services are brought onto a common compatible and unified intelligence sharing platform in terms of hardware, software, protocols and procedures. This also has to be compatible with non-military intelligence agencies and expands beyond the military realm. Lastly, when it comes to defensive IW operations, having independent service specific networks may appear to be beneficial because they are isolated but the resultant penalties are a curse. Robustness has to be achieved through multiple layers of communication and redundancy, not isolation. Joint operations are a necessity of modern warfare and it cannot be achieved without a seamless network in place. It also implies that defensive IW operations (spectrum management, cyber defense, encryption, and threat detection/mitigation) must be consolidated. It can be achieved by having common networks, equipment and software and possibly also a specialist inter-service communications branch like the medical corps. Unified theatre commands simplify many of these problems, but the IW force structures and doctrine can still be tailor made to optimize the Indian IW capability.

- **Domestic Political Considerations**. The Chinese PLA is an adjunct of the ruling CPC and the Pakistan Army is the kingmaker. The US despite being a democracy similar to India has a very different military-bureaucratic culture. The Indian situation is entirely different with the Indian Armed Forces enjoying much lesser political and bureaucratic leverage (and trust) than any of the aforementioned countries. Under these circumstances, it is not possible for the Indian Armed Forces (or any other ministry or department for that matter) to have a dedicated organization with exceptional and sweeping national and global surveillance authority like the US NSA or the PLA Joint/General Staff Department. Yet, such capability is essential if cross-border IW campaigns have to be thwarted and operations against enemy information infrastructure and space based assets are to be launched in synergy with kinetic means during war. Intelligence agencies do not meet this requirement because they only conduct acquisition operations and not attack or defense. Acquisition, attack and defense operations are continuous and complementary and it requires a dedicated force structure independent of the intelligence capability. The only viable solution is to have more trust and allow the Armed Forces to have dedicated IW organizations like the US NSA (offensive) and the US Cyber Command (defensive). Similar procedural checks and balances can be put in place to avoid overstepping of mandate. Reforms in the organization of the Indian Ministry of Defense and creation of unified commands will help.

- **Doctrinal and Geostrategic Factors.** Unlike the US which projects power globally through large expeditionary forces, or China which aspires to do so in future to counter the US influence, the Indian doctrine is essentially one of credible deterrence. The Indian Armed forces are thus mandated to preserve territorial integrity and counter any threats in its immediate neighborhood. Offensive operations are only meant to punish enemy adventures and not to annex territory per se. When it comes to IW, this translates into a credible capability to not only defend own networks but also deter adversaries through a credible capability capable of causing crippling damage to the enemy's command and control networks and critical infrastructure in case of any attack. The focus therefore has to be on electromagnetic spectrum dominance in area of operations, SEAD aircraft, airborne IW assets like UAVs and AWACS, cyber warfare against enemy critical infrastructure, non-kinetic denial of space and above all comprehensive integration of IW capability with kinetic means. India's geostrategic advantage of being able to dominate sea lanes of communication in its neighborhood must extend to surveillance, intelligence and spectrum. It is only the maritime domain where India and China do not share a common border and the Island territories offer excellent outposts which must be leveraged as IW force multipliers. In terms of defensive capabilities; India shares its land and air borders with its most likely potential adversaries and therefore the reliance on space based assets for command and control is relatively lesser and can be offset through terrestrial networks. The defensive IW effort therefore involves extending a redundant non-radiating and high bandwidth communication backbone all across the borders and also creating a secure and robust tactical mobile communication grid for maneuver elements operating on land, sea and air. This requires having a secure military satellite communication overlay having footprints covering the entire area of operations. Lastly, an indigenous space based PNT system is important though extensive reliance on the same is increasingly fraught with risk as these satellites will most likely be denied during future wars.

- **Indigenous Technology**. Doctrines and force structures remain hypothetical in the absence of equipment and in case of IW it is both hardware and software. Modern warfare is a battle of technological superiority and the decisive technologies are not for sale. Know-how related to cyber warfare, EW, DEWs, and encryption are one of the most closely guarded secrets and the only long term solution is to create indigenous capability. DEWs in particular have the potential to completely transform warfighting and deserve special attention. AI based offensive and defensive systems are yet another transformational domain. Once indigenous capability is gradually created, force structures and doctrinal evolution automatically follow suit but it requires a sustained national effort. A case in point is how new missile units have got raised once the Prithvi and Agni ballistic missile systems got perfected. India has to urgently address this shortcoming. The IEWP not only requires reviving; but also an expansion into an Integrated IW program (IIWP) encompassing not just EW but also cyber warfare, DEWs, non-kinetic ASAT and AI. If there is one area where the "Make in India" defense initiative must focus, it is this. The combined involvement of defense R&D labs, the Indian Space Research Organization (ISRO) and industry is essential, and it has to be steered as a special project as was intended when the IEWP was first instituted. The abandonment of IEWP remains the most tragic chapter in India's defense modernization effort. In the interim, the only choice is to judiciously allocate and group resources in a manner that operational payoffs are maximized. Importing hardware and software to fill critical gaps is also essential. State-of-art technology will never be offered, but some capability is better than nothing.

- **Terrain**. Terrain modulates all IW capabilities. When we speak of terrain, it includes both the "physical terrain" as well as the "information terrain" which we have discussed in this book. From the Indian context these manifest as follows:
  - *Physical Terrain.* The northern land border India shares with China and Pakistan comprises of mountainous terrain which favors defense but comprises of extremely harsh, sparsely populated and inhospitable regions many of which are inaccessible in winters. The deployment space is limited and the axes of attack and logistics are treacherous. Airborne IW systems are more effective in such terrain and ground based systems have to be specially designed for mountainous terrain and extreme weather conditions. The same applies to defensive IW systems as well. When it comes to the western borders with Pakistan, it comprises of plains as well as deserts and riverine terrain. The terrain is much more conducive to offensive operations and maneuver based warfare. Intimate IW support to mechanized formations is critical and calls for platforms to have matching cross country mobility with forces they are supporting duly augmented by airborne systems mounted on UAVs, fixed wing aircrafts and helicopters. The same applies to defensive IW systems which must be able to project deep into enemy territory. The island territories likewise are unique and require amphibious offensive and defensive IW systems including a secure satellite communication overlay. All IW capabilities require linguistic experts who are fluent in the many diverse languages and dialects spoken in the operational areas. Indian IW force structures must cater to all these.
  - *Information Terrain.* India unlike China does not exercise intimate control over information flowing in and out of the country over

the Internet. The Indian media enjoys great freedom of expression but is increasingly biased due to corporate ownership and commercial interests. The use of English as the common binding language of governance and business amplifies western influences and opinions at the cost of views from non-English speaking nations like Russia and China. Conversely it also gives India a greater leverage to influence global opinion. Overall the lack of intimate Internet censorship and being part of the English language ecosystem makes India much more vulnerable to external cognitive and cyber warfare operations. The increasing focus on Internet governance and integrating all databases with the biometric Aadhaar system run by Unique Identification Authority of India (UIDAI) similarly requires an extremely robust defensive IW capability. Protection of critical infrastructure against cyber attacks is another extremely specialized domain and this has to be steered by a multi-disciplinary organization at the national level. There are no major force structure related lessons here because Internet, media and cyberspace are not pure military domains. However, the IW plans at the strategic and operational levels have to be alive to these realities and be prepared to attack as well as defend in cyberspace. In particular, India needs to create organizations to isolate, analyze and control Internet information in case of external attack without compromising on freedom of speech. It is a delicate balance, yet achievable without descending into a surveillance state. Similarly organizations to counter propaganda and legal warfare are a must and these must include the military. The Armed Forces are a vital stakeholder during war and CI and must be adequately represented in all such organizations even though they may not be suitable to be in the leadership role particularly in case of operations involving critical infrastructure. Creation of Territorial Army IW Battalions is an idea whose time has come.

- **Space Capabilities**. India has reasonable space prowess but its space program is not military oriented. India does not have a dedicated military satellite network like the US or any credible kinetic counter space capability like what China is creating. This has major IW related connotations because the lack of counter-space capability enables adversaries to leverage their own assets to gain asymmetric advantage while stymieing our own. The Indian armed forces as a result shall be denied of not only critical surveillance and PNT capability but also the ability to project forces much beyond its land borders. Also, the use of commercial satellites for military purposes leads to false confidence – these are not adequately hardened against attacks and will be neutralized the day war starts with catastrophic consequences. If not a dedicated military space program and kinetic ASAT weapons, at least a credible EW and cyber warfare capability to deny space based assets to adversaries must be urgently created to have some deterrence.

Force Structures for India

The Indian Armed Forces are striving to modernize and re-structure but are hampered by lack of indigenous equipment manufacturing capability and inadequate political understanding about how warfare is changing. Even within the Armed Forces there is resistance to break the status-quo and unify. The false belief that strategy and tactics of the past combined with diplomacy, leadership and valour of foot soldiers fighting in the land, sea and air battlespaces will result in victory still persists. Failure to occupy the new battlespaces of information and space where there are no boots on ground is however a recipe for disaster. A decade or two from now there may be no men behind the machines and the effort to transform has to begin now; else the gap in capabilities versus other modern armed forces will rapidly grow into a chasm with disastrous consequences.

Some recommendations are given below.

- The Indian political compulsions will not allow for a global offensive military IW organization but a defensive organization (like the US Cyber Command) to protect the military portion of the information battlespace is an equally important necessity and would be readily acceptable. It would be appropriate to call it the Indian Network Command (or something similar) instead of Cyber Command which is misleading. Such an organization however calls for the communication elements of army, navy and air force to be unified and reorganized on the lines of the medical corps wherein they continue to provide intimate support to respective services but yet have a common higher hierarchy which steers the development of new systems and creates a unified global communication, encryption and cyber defense capability similar to what the US Cyber Command does. It also lays the bedrock for creation of unified theatre commands in the future which are a necessity of modern warfare. It maybe given an offensive mandate in future if enough confidence exists.

- The ability to dominate electromagnetic spectrum is the most crucial element of future warfare. Spectrum is the life blood of the network centric battlefield and all modern and futuristic weapon systems heavily rely on uninterrupted availability of electromagnetic spectrum. Denial of spectrum therefore has the ability to completely nullify technological superiority. The reason why the most technologically sophisticated US Navy recently reintroduced celestial navigation as part of its training is because it realizes that its adversaries have developed asymmetric EW capabilities that will deny GPS[8]. Essentially it is an acknowledgement that EW can push a modern force back into the pre-NCW era tactics. The manner in which Russia has used EW to target state of art US systems in Syria and Ukraine carries valuable lessons in how spectrum can be leveraged to create asymmetric advantages[9,10,11]. China is following a similar path. There is no option but to urgently revive the IEWP and treat it as a highest priority national project to be steered by high level inter-ministerial committees with Indian industry (particularly IT and Electronics), armed forces, academic institutions, defense labs and ISRO working in partnership as was originally envisaged. The original IEWP must however be expanded into a national Integrated IW Program (including cyber warfare and DEWs) and the new effort should be delinked from the defense procurement procedures. The program must also include other IW stakeholders like the National Technical Research Organisation (NTRO), Intelligence Bureau (IB) and the Research and Analysis Wing (R&AW). Such a program can not only transform the Indian IW capability but also provide a tremendous fillip to the indigenous high-tech manufacturing capabilities in the long run. In the interim, gaps in capability must be filled through imports. New EW units to fill the voids must be raised and grouped with tactical formations. Joint EW staff is likewise required to be created for inter-service coordination. Lastly a specialized tri-services EW cum cyber warfare unit for non-kinetic counter space operations must be created as a special force.

- IW capabilities must integrate with kinetic means and it involves the army, navy as well as air force. It requires integration at hardware and software level and is not easy to achieve because India imports most of its weapon systems and they have proprietary design. The lack of unified commands further complicates such integration. Regardless of hurdles, this has to be made a priority and a solution found through ad-hoc interim organizations, custom built indigenous software and hardware overlays as well as field trials to validate the integration.
- HR policies and systems must be modified to induct and retain technically specialized manpower. Creation of IW TA Battalions is one possible method to attract highly qualified specialists from the IT and telecommunication industry who otherwise may not be interested in a permanent military career. Creation of state-of-art joint training infrastructure to include IW training ranges is a must.
- Offensive cyber warfare capabilities must be created and India must have its own tailor made equivalent of the US NSA and China's GSD. However, both the US and Chinese models are inappropriate in the Indian context. Also the Internet is not an exclusive military domain and the highly specialized nature of cyber warfare and critical infrastructure requires a joint multi-disciplinary cyberspace organization at the national level with a tri-service military vertical as a sub-specialty. The Armed forces must however collectively push for creation of such an organization because operations against enemy critical infrastructure (and defense of own) are now very much a part of modern warfare. Existing organizations like NTRO can be leveraged. A dedicated military vertical is a must in any such organization because large scale attacks on critical infrastructure would invariably happen in a warlike situation and require integration with military plans, kinetic means and the military IW capabilities.
- Psy warfare has been a neglected discipline and Indian Armed Forces do not have dedicated psy warfare units. Cognitive operations are becoming increasingly important and either new psy warfare units must be raised or else some of the existing units need to be dedicated to this role. Since cognitive perceptions cannot be directly manipulated, the actual operations will be conducted in the physical and electronic (Internet) terrains. The units have to be accordingly designed. It is

also important that perception management and public relations be treated as a part of IW operations.

- Legal warfare is an increasing threat and requires a dedicated organization at the national level. Once again the military cannot be in the lead role in any such organization due to lack of expertise in international law and diplomacy but they will be an essential vertical since the disputes will often involve border areas whose sanctity and integrity is maintained and monitored by the armed forces. The military IW staff at strategic and operational level should add such a role to their charter.
- The Indian Army has experimented with the concept of IW Brigades. This was necessary and brought home important lessons. As already explained; any grouping of IW capabilities has to be synergistic and combined only if they reinforce each other. Cyber warfare and SIGINT is one such combination applicable at the strategic level but merging psy warfare, EW and cyber warfare at the tactical level is not necessarily good. The level at which capabilities are being combined (strategic, operational or tactical) is also important. These lessons must be borne in mind as the Indian Armed Forces strive towards more modern force structures and operational doctrine.

ENDNOTES

1. Pillarisetti, Jagan, "*Descent Into Danger – The Jaffna University Helidrop*" (Published on the Bharat Rakshak website on 24 November 2009), Web URL: http://www.bharat-rakshak.com/IAF/History/1987IPKF/1024-Chapter03.html accessed on 15 May 2018
2. Carlo Copp, "*Operation Desert Storm – The Electronic Battle: Parts 1-3*" (Published on Air Power Australia website from June to August 1993), Web URL: http://ausairpower.net/Analysis-ODS-EW.html accessed on 20 January 2017
3. Srivastava, Ishan, *"How Kargil spurred India to design own GPS"*, Published on the Times of India website on 05 April 2014), Web URL: https://timesofindia.indiatimes.com/home/science/How-Kargil-spurred-India-to-design-own-GPS/articleshow/33254691.cms accessed on 15 May 2018
4. Gupta Shishir, "*99 phone tapes show General kept Sharif in dark on Kargil, in book he says opposite*", (The Indian Express, 26 October 2006)
5. "*Samyukta presented to the Army*", (The Hindu, 20 January 2004)
6. Electronic Warfare professionals are referred to as "*Crows*". The Association of Old Crows is an organization for individuals who have common interests in Electronic Warfare (EW), Electromagnetic Spectrum Management Operations (EMSO), Cyber Electromagnetic Activities (CEMA), Information Operations (IO), and other information related capabilities.
7. Pandit, Rajat, "*Army gears up to shed flab, turn lean, mean & fast*", (Published on the Times of

India website on 20 August 2018), Web URL: https://timesofindia.indiatimes.com/india/army-plans-transformation-drill-to-be-lean-and-mean-force/articleshow/65465504.cms accessed on 25 August 2018

8. Samseer M, "*US Navy reintroduces celestial navigation to counter cyber attacks*", (Published on Naval Technology website on 15 October 2015), Web URL: https://www.naval-technology.com/news/newsus-navy-reintroduces-celestial-navigation-to-counter-cyber-attacks-4695720 accessed on 20 January 2017
9. Colin Clark, "*Russia Widens EW War, 'Disabling' EC-130s OR AC-130s In Syria*", Published on Breaking Defense website on 24 April 2018), Web URL: https://breakingdefense.com/2018/04/russia-widens-ew-war-disabling-ec-130s-in-syria/accessed on 15 May 2018
10. Loren Thompson, "*Electronic Warfare: How The U.S. Army Could Lose Its Next War*", (Published on Forbes website on 15 March 2016), Web URL: https://www.forbes.com/sites/lorenthompson/2016/03/15/electronic-warfare-how-the-u-s-army-could-lose-its-next-war/#3c7580251b09 accessed on 15 May 2018
11. Col Liam Collins, "*Russia Gives Lessons in Electronic Warfare*", (Published on Association of the United States Army website on 26 July 2018), Web URL: https://www.ausa.org/articles/russia-gives-lessons-electronic-warfare accessed on 25 August 2018

Index